Arctic Pipeline Planning

Gulf Professional Publishing is an imprint of Elsevier
The Boulevard, Langford Lane, Kidlington, Oxford, OX5 1GB, UK
225 Wyman Street, Waltham, MA 02451, USA

First published 2013

Notices
Knowledge and best practice in this field are constantly changing. As new research and experience broaden our understanding, changes in research methods, professional practices, or medical treatment may become necessary.

Practitioners and researchers must always rely on their own experience and knowledge in evaluating and using any information, methods, compounds, or experiments described herein. In using such information or methods they should be mindful of their own safety and the safety of others, including parties for whom they have a professional responsibility.

To the fullest extent of the law, neither the Publisher nor the authors, contributors, or editors, assume any liability for any injury and/or damage to persons or property as a matter of products liability, negligence or otherwise, or from any use or operation of any methods, products, instructions, or ideas contained in the material herein.

British Library Cataloguing in Publication Data
A catalogue record for this book is available from the British Library

Library of Congress Cataloging-in-Publication Data
A catalog record for this book is available from the Library of Congress

ISBN: 978-0-12-416584-7

For information on all Gulf Professional Publishing
publications visit our website at **store.elsevier.com**

This book has been manufactured using Print On Demand technology. Each copy is produced to order and is limited to black ink. The online version of this book will show color figures where appropriate.

ELSEVIER · Book Aid International

Working together
to grow libraries in
developing countries

www.elsevier.com • www.bookaid.org

Transferred to Digital Printing in 2013

DEDICATION

This book is dedicated to my father, Kripashanker Singh, from whom I learned to put pen on paper, and who taught me to be analytical in my approach, and instilled a sense of discipline and dedication to any work that I do.

CONTENTS

Preface xiii

Acknowledgment xv

PART 1 HISTORY OF CONSTRUCTION IN THE ARCTIC

Chapter 1 Introduction 3

1.1 Introduction and Background to Design and Construction
of Arctic Pipeline 3

1.2 Typical Challenges 4

1.3 Current Arctic Proposals 5

1.4 Technology Transfer to Other Areas of the World 6

1.5 Offshore Pipelines in the Arctic Area 7

Chapter 2 Design and Construction 9

2.1 Introduction to the Arctic Environment 9

2.2 Addressing the Challenges of Design 10

2.3 Challenges of Designing Arctic Pipeline 10

2.4 River and Waterways Crossings 11

2.5 Calculations for Design 11

2.6 Examples of Calculation 12

2.7 Expansion and Flexibility 15

2.8 Unsupported Span of Pipe 16

2.9 Buoyancy of Pipes 17

2.10 Calculating the Stresses in the Pipe at the Transition from
Belowground to Aboveground 19

2.11 First Scenario, Reducing Stresses Without Anchor Block 20

2.12 Second Scenario, Reducing Stresses with Anchor Block 21

2.13 Strain-Based Design 22

2.14 Construction Methods 26

2.15 Protection of Tundra 26

2.16 Grade and Slopes..26

2.17 Construction in Warm Permafrost...26

2.18 Design and Construction in Thaw Settlement of
      Gas Pipeline 28

2.19 Ditching and Backfilling..29

2.20 Control Valves..29

2.21 Pumps for Crude Oil Pipelines..30

2.22 Designing and Construction to Deter Ice Gouging....................31

2.23 Design and Construction for Subsea Arctic Pipelines...............31

2.24 Design and Construction of Pipeline Crossings 32

2.25 Thermal Expansion and Contraction..33

2.26 Design of Compressors and Pump Stations 38

2.27 Design of Compressor Stations..39

2.28 Designing a Gas Compressor Unit 39

2.29 Liquid-Removal Equipment..39

2.30 Emergency Shutdown Facilities 40

2.31 Station Piping 41

PART 2 SAFETY AND COMMUNICATION

Chapter 3 Safety on Construction Sites 45

3.1 Hazards to Construction Workers...46

3.2 Applicable Safety Regulations...47

Chapter 4 Pipeline System Communication 49

4.1 System Concepts 50

4.2 The Human−Machine Interface 51

4.3 Hardware Solutions..53

4.4 Remote Terminal Unit..54

4.5 Supervisory Station 54

4.6 Communication Infrastructure and Methods 55

4.7 SCADA System for Typical Gas Pipeline 57

4.8   SCADA Operator Workstations 58

4.9   SCADA Servers 59

4.10   Selection of Software 60

4.11   Communication Infrastructure Security.................................60

4.12   Typical Codes and Standards..61

Chapter 5 Electrical Equipments 63

5.1   Introduction 63

5.2   Design of Electrical Equipment..64

5.3   Conductor Size...68

PART 3 SELECTION OF MATERIALS

Chapter 6 Introduction to the Material Section 73

**Chapter 7 Material Properties in Low Temperature
            Environment 75**

7.1   Ductility and Behavior of Steel in a Low
       Temperature Environment 75

7.2   Concept of Toughness and Loss of Toughness in a
       Low Temperature Environment 77

7.3   Fracture Toughness K_c 78

7.4   Metal Strength at Low Temperature.......................................81

7.5   Types of Impact Tests 82

7.6   Energy Absorption in Impact Testing 83

7.7   Transition Temperature for Energy Absorption.........................83

7.8   Transition Temperature for Lateral Expansion.........................84

7.9   Drop-Weight Tear Test (DWTT)..84

7.10   Selecting Material from Specification and Code Books............85

Chapter 8 Line Pipes..89

8.1   Metallurgical Considerations for Line Pipe Steel 90

8.2   Submerged Arc Welded Line Pipes 93

8.3   Classification of Line Pipes 94

8.4 PSL 1 vs. PSL 2 96

8.5 Determination of Percentage Shear Through DWT Test
     for PSL 2 Welded Pipes...96

8.6 Ordering a Line Pipe 97

8.7 Pipes from Other Specifications 97

Chapter 9 Fittings and Forgings 99

9.1   ASME/ANSI B16.5—Pipe Flanges and Flanged Fittings........ 102

9.2   ASME/ANSI B16.9—Factory-Made Wrought Steel
      Butt Welding Fittings... 103

9.3   ASME/ANSI B16.11—Forged Steel Fittings, Socket
      Welding and Threaded Connections 104

9.4   ASME/ANSI B16.14—Ferrous Pipe Plugs, Bushings and
      Locknuts with Pipe Threads... 104

9.5   ASME/ANSI B16.20—Metallic Gaskets for Pipe Flanges
      Ring-Joint, Spiral-Wound, and Jacketed.......................... 104

9.6   ASME/ANSI B16.21—Nonmetallic Flat Gaskets for
      Pipe Flanges 105

9.7   ASME/ANSI B16.25—Butt Welding Ends 105

9.8   ASME/ANSI B16.28—Wrought Steel Butt Welding
      Short Radius Elbows and Returns 105

9.9   ASME/ANSI B16.36—Orifice Flanges 106

9.10 ASME/ANSI B16.39—Malleable Iron Threaded
      Pipe Unions 106

9.11 ASME/ANSI B16.42—Ductile Iron Pipe Flanges and
      Flanged Fittings, Classes 150 and 300 106

9.12 ASME/ANSI B16.47—Large Diameter Steel Flanges:
      NPS 26 through NPS 60 107

9.13 ASME/ANSI B16.48—Steel Line Blanks 107

9.14 ASME/ANSI B16.49—Factory-Made Wrought Steel
      Butt Welding Induction Bends for Transportation
      and Distribution Systems.. 107

Chapter 10 Valves **109**

10.1   API 598 Valve Inspection and Testing 109

10.2   API 600 Steel Valves—Flanged and Butt Welded Ends.........109

10.3   API 602: Compact Steel Gate Valves—Flanged, Threaded,
        Welding, and Extended Body Ends 110

10.4   API 603 Class 150 Cast, Corrosion Resistant Flanged
        End Gate Valves..110

10.5   API 608 Metal Ball Valves—Flanged and
        Butt Welded Ends 110

10.6   API 609: Butterfly Valves—Double Flanged, Lug
        and Wafer Type...111

10.7   Testing of Valves 111

10.8   NACE Trim and NACE Material..111

10.9   ASME Codes and Other Specifications...............................112

10.10   API 6D Specification for Pipeline Valves.............................114

10.11   API 6FA 115

10.12   API 526 Flanged Steel Pressure Relief Valves.....................116

10.13   API 527 Seat Tightness of Pressure Relief Valves (2002).......116

10.14   ANSI/API STD 594 Check Valves: Flanged, Lug,
        Wafer, and Butt Welding 116

10.15   ANSI/API 599: Metal Plug Valves—Flanged, Threaded,
        and Welding Ends..116

10.16   ASME/ANSI B16.38—Large Metallic Valves for Gas
        Distribution...116

10.17   ASME/ANSI B16.33—Manually Operated Metallic
        Gas Valves for Use in Gas Piping Systems Up to 125 psig......117

10.18   ASME/ANSI B16.40—Manually Operated
        Thermoplastic Gas Valves..117

10.19   ASME/ANSI B16.10—Face-to-Face and End-to-End
        Dimensions of Valves...118

10.20   ASME/ANSI B16.34—Valves—Flanged, Threaded,
        and Welding End 118

Index **119**

PREFACE

Writing a book on a dry subject such as arctic pipeline, on which not much is available for reference, is a challenge. I was also challenged by the fact that there is not much difference in the primary approach of pipeline design and construction of arctic pipeline and that of a pipeline on land and sea in warmer conditions. The only analogy that comes to my mind is, if the canvas was suddenly pulled away from Picasso and he was asked to paint on say, a clay tablet, he would still be painting and painting beautifully but the demands of the new media would be challenging. Designing and subsequently constructing a pipeline in an arctic environment is somewhat like painting on a very different canvas—the environmental setting.

This book is aimed at those managers, engineers, and non-engineers who are looking for some insight into what is required to design and construct pipeline in arctic regions. It is not intended to be a step by step, how to do detailed procedure of design and construction. In the absence of specific parameters, this book cannot be used to design a pipeline; it is only intended to be an introduction to the subject. After the introduction, the reader is expected to go out and experience the Arctic.

Those individuals who need more detailed study on any specific topic covered in this book, must reach out to the acknowledged specialized associations, institutions, and local regulatory bodies for further guidance. There are several published works available from these bodies that can be of immense help in developing in-depth understanding of specific subjects.

ACKNOWLEDGMENT

Writing this book made me realize how dependent on outside agencies of various types a person is in accomplishing a task of this type. It also made me realize the power of nature which we try to challenge and overturn with construction technologies and engineering expertise but end up making a compromise with. A compromise in which nature is always the winner—and by letting nature win, engineering excels.

I am extremely grateful to the management and team of Gulf Interstate Engineering, Houston (www.gie.com) for creating an environment that encouraged me to write this book. I am also grateful to the clients of Nadoi Management, Inc. who allowed me to use data and pictures from the work that I performed for them as consultant.

I am also indebted to the support and help from my friend Olga Ostrovsky. She helped me negotiate the obstacles of writing and editing the drafts. Without her expert help this book would not have been possible.

Last but not the least, I am also grateful to my loving wife Mithilesh and my son Sitanshu for their support in accomplishing this goal. Their support and understanding has been immense during this work.

PART 1

History of Construction in the Arctic

CHAPTER *1*

Introduction

1.1 INTRODUCTION AND BACKGROUND TO DESIGN AND CONSTRUCTION OF ARCTIC PIPELINE

Arctic pipelines are trunklines transporting gas and oil from production sites to the facilities that process the product and the pipelines that convey it to the consumers. The pipelines that are constructed in the territory limited within the polar circle are called arctic pipelines. The differentiation of pipelines into arctic and those which are located to the south of the polar circle is notional. The main characteristics of arctic pipelines are determined by the climatic conditions pertaining to a given territory, which are described below, and it may be noted that some of these features can be encountered in territories outside of the polar circle.

- Low radiation balance
- Summer average zero temperature
- Annual average subzero temperature
- Large areas of permafrost
- Wet lands and an abundance of swamps in plains.

The Arctic is beautiful and environmentally vulnerable, and at the same time has huge resources that society will want to take advantage of. Oil companies are nowadays highly aware of the possibility of environmental damage, extremely sensitive about it, and firmly supervised. There is no lack of challenging and unexpected problems for the pipeline engineer.

Pipeline construction in the Arctic was not very much a considered option until the 1300 km Trans-Alaska Pipeline was proposed about 40 years ago. Following BP's discovery of 25 Bbbl of oil in Prudhoe Bay oil field near the Arctic Ocean coast in 1968, the Trans-Alaska Pipeline was proposed. Prior to the construction of the Trans-Alaska Pipeline, only pipelines of short lengths had been constructed in places such as Siberia and close to Norman Wells on the Mackenzie River, Canada.

As a precursor to the Trans-Alaska Pipeline project a demonstration project took the tanker Manhattan through the North-West Passage in 1969, which showed that it was feasible to make the voyage in the summer, but that a year-round transportation route would be difficult if not impracticable. The alternative to the above was a 48 in. diameter pipeline running north to south for 1300 km from Pump Station-1 near Prudhoe Bay on the Beaufort Sea to the ice-free port of Valdez, southeast of Anchorage. Thus, the Alaska pipeline was conceived. The pipeline had a total 12 pump stations, of which the first 5 were located before the Yukon Crossings, the last was at Thompson pass. An approximately 200 km long gas pipeline ran parallel to the oil pipeline from Prudhoe Bay to Dietrich Pass.

In its time, the project was high profile and hugely controversial bearing in mind the environmental impact of the pipeline; fear was expressed that the project would "ruin Alaska". That idea is still not fully settled conclusively. The permits were issued for the project after prolonged hearings and lawsuits. The construction began in 1974 and was completed at a cost of $8 billion. The operations began in 1977.

1.2 TYPICAL CHALLENGES

A land pipeline in the Arctic runs into potential technical difficulties. Like the majority of geotechnical problems, most of these difficulties are linked to the flow of water and heat.

Some of the distinguishing challenges to an arctic pipeline relate to the unique composition of the land and soil in which the pipe is laid. Frozen soil is very strong and can bear large loads. On the other hand, if the fluid in the pipeline is hot and this heat is conducted through the pipe walls into the soil, the frozen soil would thaw. If the thawing soil has a large ice content, it will turn into soft mud; this soft soil or mud no longer has same strength as the frozen soil and cannot carry the same load – in fact it can hardly support any load and the result would be the section of pipeline so affected sinking. The amount of ice in frozen soil varies enormously; this variation is not uniformly distributed even over quite short horizontal distances, and therefore the amount of sinking of a pipeline would also vary. As a result of a sinking pipeline in thawing soil, a pipeline might become severely bent, and in extreme cases might even buckle and crack.

This problem can be addressed by insulating the pipeline and by raising the pipeline above the frozen ground on piles. A majority of the sections of arctic pipelines are constructed on pile supports for this and various other reasons, which are discussed further in this book.

At the time of the conceptual engineering of the Trans-Alaska Pipeline, pessimists thought that it might be necessary to do this for about 60 miles or so, but in the end a much greater length was raised above the ground on piles. That solution created other difficulties, some natural and some not so natural and which are man made. Reindeer or Caribou migrate, and it was opined that they would see a raised pipeline as a barrier, but in reality they turned out to be more intelligent than those experts gave them credit for. They are intelligent enough to walk underneath the raised sections of pipeline. The man-made difficulties pertaining to a raised pipeline include accidents and damage associated with guns and hunting, which is common in Alaska. This combination leads to some accidental, or sometimes deliberate, shooting at the pipeline.

The Trans-Alaska Pipeline has now been in operation for about 40 years, without any major incident to the main line. The main exception is the oil spill in 2006 of about 6000 bbl from a flow line. Over the pipeline's lifetime it has transported over 15 Bbbl of oil. The throughput has now decreased, from more than 2 MMbbl/d at the peak to around 1 MMbbl/d now, and some of the original eight pump stations have since been decommissioned.

1.3 CURRENT ARCTIC PROPOSALS

The success of the Alaska pipeline has opened ways to develop more pipeline projects in arctic regions. Approximately 30 trillion cubic feet of stranded gas lies under the North Slope of Alaska that needs to be tapped and brought to market, and for this purpose a pipeline is being proposed. Many different competing schemes have been investigated, but the most likely to succeed is a pipeline running south to the Fairbanks area, and then southeast along the Alaska Highway into Canada. An unlikely alternative is an offshore pipeline in the Arctic Ocean, eastward parallel to the coast, to link into a Canadian gas pipeline from the Mackenzie Delta region.

The fate of the Mackenzie Valley Gas Pipeline has been argued about for almost as long as the Trans-Alaska Pipeline has existed but nothing has been built. Much engineering was done in the 1970s, and there were hearings dealt with by the Berger Commission, but in the end the Commission ruled against the project, essentially on socioeconomic grounds rather than technical.

Alternative projects have invested large sums in site investigation and engineering. The most recent lead project is the Mackenzie Gas Project, led by ExxonMobil and its partners, and there were public hearings in Yellowknife in 2006. Some First Nations aboriginal groups are partners in the project; however, others are opposed. The project appears currently to be "on the back burner," in part because the estimated cost suddenly leapt from $9 billion to $16 billion.

1.4 TECHNOLOGY TRANSFER TO OTHER AREAS OF THE WORLD

One benefit of all these pioneering projects is that they throw up fresh ideas and add experience to be used in a different environment – some of them directly, some morphing into newer concepts. The experience and technologies developed may become more helpful to regions that are south of the arctic regions but face somewhat similar environmental challenges; for example projects in northern Canada and the Caspian Sea regions are to the point. Another such project that might derive some benefits is the proposed $US7.6 billion proposed gas pipeline from Yoloten-Osman fields of Turkmenistan, Afghanistan, Pakistan, and India (TAPI). The pipeline would traverse the unforgiving mountainous terrains of Turkmenistan and northern Afghanistan, which are extremely cold, and encounter many different soil conditions though not permafrost as in arctic regions but still challenging nonetheless, and also environmentally sensitive as in arctic regions. This pipeline is estimated to carry about 90 million standard cubic meters per day (mmscmd) of gas, of which 14 mmscmd would be bought by Afghanistan and 38 mmscmd would each be for India and Pakistan. The 1735-km long pipeline will run from Turkmenistan's Yoloten-Osman gas field to Herat and Kandahar province of Afghanistan, before entering Pakistan through Quetta and reach Multan before ending at Fazilka in India.

1.5 OFFSHORE PIPELINES IN THE ARCTIC AREA

There have been several pipelines established in frozen seas, especially in the Caspian Sea and some in seas around the Russian tundra.

The proposed oil exploration by Royal Dutch Shell in cusps of the Arctic Ocean, Beaufort Sea and Chuchki Sea about 100 miles offshore and northwest of Barrow, the northernmost municipality in USA, is another new frontier in oil exploration in the Arctic. It is estimated that this field has about 27 billion barrels of recoverable oil. This will be the next big project. The drilling permits have been issued, and operations are likely to start sometime in 2012. Shell plans to drill oil wells to produce oil. The drilling is planned to a depth of about 150−200 ft. This is shallow as compared with over 5000−10000 ft depths found in the Gulf of Mexico and in that respect it is less challenging. But that is only one aspect; the water may be shallow but the challenges of facing crushing ice floes, arctic hurricanes and total darkness for 70 days of the year are typical. Any oil spilled under the ice will be near impossible to deal with. Environmentally too it can have serious repercussions as the drilling location overlaps with migration routes of whales and other marine lives. In the event of any spillout, there must be plans to clean up the spill or to blowout if that ever happens.

Shell, or indeed any other oil company, cannot guarantee accident-free operations, so this fact must be acknowledged and associated safety measures must be planned into the project schedule.

On the basis of all information available to the public, Shell has learned from the failure of Deepwater Horizon in the Gulf of Mexico and plans to use the blowout protector that will have an extra cutoff valve just in case the primary valve fails. It is learned that Shell is also likely to keep a standby capping device similar to the one that was used to cap the Deepwater blowout. The equipment will be placed on the icebreaker support ship positioned between Chuchki and Beaufort drilling locations.

CHAPTER 2

Design and Construction

2.1 INTRODUCTION TO THE ARCTIC ENVIRONMENT

The challenge of construction and maintenance is more serious in the arctic because of the severe elements of nature. The working conditions and the terrain are very unforgiving. For example, the temperature often reaches $-74°F$ ($-8.8°C$) windchill. The extreme cold limits the ability of man and machine. Movement is impaired and mental activity is reduced, raising the possibility of mistakes leading to accidents and mishaps which could cause injuries and even death.

In the arctic environment, the weather can vary significantly from area to area, especially in the winter. The vagaries of the weather can pose significant challenges to construction efforts and can test the endurance of men, machines, and material. This is true even in the arctic summer.

In arctic regions, a lot depends on whether the work site is close to the Arctic Ocean or further inland. In general, sites farther inland are colder in the winter and warmer in the summer. In my experience, typical inland temperatures in the winter range from $-25°F$ to $+10°F$; in the summer the temperature in degrees Fahrenheit ranges from the mid 30s to the mid 50s.

Summer or winter, windchill can have a huge impact on the comfort factor. This demands that workers dress appropriately keeping in mind that weather conditions can change very quickly and that they are prepared for the worst, keeping a constant eye on the posted forecast.

The warmer temperature of summer brings with it another peril of the nature, the mosquitoes and bugs; they thrive in warmer weather, and they are not very human friendly.

Added to the above are the other natural challenges such as the presence of wild animals and the consciousness of disturbing their natural habitat as little as possible. The pipeline must not infringe on the day-to-day lives of wildlife and coexistence is the watchword.

2.2 ADDRESSING THE CHALLENGES OF DESIGN

Pipeline and facilities in arctic regions are designed in accordance with relevant industry specifications or codes such as ASME/API/CSA and additional requirements imposed by the regulatory authorities. There are few fundamental differences when designing pipelines for carrying oil or gas. This is primarily due to the properties of the two fluids and how they behave in different environments. The discussion here is primarily of general nature; however, wherever it is possible, the differences in design practices will be identified.

Some of the design calculations discussed below are in nature the same as they would be for any other geographical setup of pipeline. However, the conditions in arctic regions are very different from conditions found elsewhere.

2.3 CHALLENGES OF DESIGNING ARCTIC PIPELINE

One of the key issues with arctic design and construction is how to address the challenges of frost heave and thaw settlement. A pipeline exposed to either frost heave or thaw settlement could experience large bending forces that typically create large longitudinal strain. The standard wall thickness determined using pressure as a factor is usually not supportive of such bending strains.

The factors that need to be considered to mitigate bending strains are as follows:

- Insulation of the pipe to limit the extent of thaw
- Construction of pipeline in an aboveground embankment to limit extent of thawing
- Deeper burial for more stable soil
- Placement of pipe aboveground to isolate the pipeline from unstable soils
- Over excavation to replace unstable soil with stable soil such as sand to reduce the magnitude of settlement
- Increasing the wall thickness of the pipeline to accommodate larger strain caused by different ground movements.

Selection of any one or a combination of the above options can be made. The increase of wall thickness is technically the most reliable option.

The anticipated maximum settlement of the pipe in thawing soils is estimated to be in the range of 1−2 m. Increasing the wall thickness of the pipe to accommodate the settlement strain does not affect the normal construction operation or procedures.

Other options have their limitations and are useful in only specific conditions in a limited way.

- The pipe insulation method is feasible and has been used in several earlier construction projects. However, this has the potential of corrosion occurring under the insulation, and deterioration of the coating limits its application.
- The aboveground embankment option is also used extensively. However, its usefulness is limited to very specific locations since the attendant maintenance costs are significantly high.
- Deep burial is possible where the thaw is shallow and soil is stable. Detailed field survey would be required to identify those locations and plan construction in those areas only. This can increase the cost and time of construction.
- The option of over excavation is limited by the degree of thaw. A large area of thaw would require removal and replacement of large quantities of soil.

2.4 RIVER AND WATERWAY CROSSINGS

Further challenges are presented by rivers and waterways crossing the route of the pipeline. A detailed hydrological, geological, and topographical survey is required. Banks of rivers often contain permafrost; the river beds may also contain permafrost if the depth of water is less than about 2 m. This will require careful consideration of design and construction planning. If the pipeline is operating at temperatures high enough to thaw the soil surrounding the pipe and so cause significant settlement, then the use of insulation to minimize heat transfer from pipe to soil may be the preferred method for mitigating thaw settlement.

2.5 CALCULATIONS FOR DESIGN

In this section, we will discuss and give calculated examples of some of the key challenges that face a pipeline engineer and design team.

The design of a pipeline in arctic conditions follows the same basic criteria as for the warmer parts of the world; for example, the pressure determination for steel pipeline is calculated using the following relationship. Three different scenarios are discussed to determine the pressure for pipelines.

$$P = (2St/D) \times F \times E \times T \tag{2.1}$$

where

P = design pressure (psi gauge; or kPa in metric units)
S = yield strength of steel (psi; or kPa in metric units)
D = nominal outside diameter of the pipe (in. or mm)
t = nominal wall thickness of the pipe (in. or mm)
F = design factor
E = longitudinal joint factor
T = temperature derating factor (if) applicable.

2.6 EXAMPLES OF CALCULATION

1. The pressure limit of a SAW (submerged arc welding) welded API 5L Grade X60 pipe of 36-in. diameter with wall thickness of 0.500 in. operating at a temperature lower than 250°F in class location class 1, division 2 where the design factor is 0.72, can be determined as follows:

   Substituting these values into Eq. (2.1) we have

   $P = (2 \times 60{,}000/36) \times 0.72 \times 1 \times 1$ (both the longitudinal weld factor and temperature derating factors are 1 in this case).

   $P = 2400$ psi.

2. If the above pipe is operating at temperatures above 250°F, say at 299°F, then the temperature derating factor will be 0.967. The calculation based on Eq. (2.1) is as follows:

   $P = (2 \times 60{,}000/36) \times 0.72 \times 1 \times 0.967$ (note the change in the value of temperature derating factors in this case).

   $P = 2320$ psi.

3. In the first example above, if the gas pipeline passes through an industrial area of a city where human occupancy is high due, say, to multistory buildings with heavy and dense traffic, or it crosses

underground utilities such that the class location is defined as class 4, in such conditions, the design factor would change to 0.40. The pressure calculation is carried out using Eq. (2.1).

Substituting the values into Eq. (2.1) we have

$P = (2 \times 60,000/36) \times 0.4 \times 1 \times 1$ (both the longitudinal weld factor and temperature derating factors are 1 in this case).

$P = 1333$ psi.

For all practical purposes, the welding processes used for making API 5L pipes have a longitudinal weld joint factor of 1 except for furnace butt welded pipes, which have a factor of 0.60. However, pipes made by use of the electric resistance welding (ERW) process to other specifications such as ASTM A134, ASTM A139, ASTM A211, ASTM A671,and ASTM A672 (class 13, 23, 33, 43, and 53) have design factors reduced to 0.80. The pipes manufactured to these ASTM specifications are not used for pipeline construction; however, they may be found in some limited applications of station piping systems.

The location class of the pipeline is an important variable to always take note of; the location class is subject to changes as the urban population changes. What could have been a class 1 location at the time of initial construction with less than 10 buildings (wasteland, desert, sparsely populated areas, etc.) may change, due to a newly thriving population density area, thus altering both the class and design factors.

The key factor for the design of arctic pipeline would be the lower temperature and the properties and behavior of the materials under such conditions. The property of materials in the lower temperature range is discussed in detail in Chapter 2 of Part 3 of this book. In this part, we concentrate on the challenges that affect other design aspects of the pipeline.

The changing temperature can cause stresses that are normally not experienced in warmer-area pipelines. The ability of a material to bear stress is the key point to understand; this is discussed further in the book.

The same relationship can also be used to calculate the required wall thickness of a pipe, if all other variables are known.

For example, if a 30-in. diameter API 5L X42 ERW or HFW pipe is installed to operate at 1480 psi pressure in a class 1 location with a design factor of 0.72, and the operating temperature is less than 250°F, the required wall thickness of this pipe can be determined using the same formula (Eq. (2.1)).

●●● ───

Example

Consider Eq. (2.1), which is repeated here.

$$P = (2st/D) \times F \times E \times T$$

We can rearrange this formula as follows:

$$PD/2sFET = t \tag{2.2}$$

Substituting the values in Eq. (2.2) we have

$$1480 \times 30/(2 \times 42,000 \times 0.72 \times 1 \times 1) = t$$
$$t = 0.734 \text{ in.}$$

Thus, an X42 grade of pipe with 0.734 in. wall thickness would suffice for the design maximum allowable operating pressure (MAOP).

Sometimes, a need arises to replace an existing section of pipe with another. Engineers are often required to take a decision on substituting one set of pipe with another pipe. Taking the above example of X42 pipe (specified minimum yield strength (SMYS) 42,000 psi) as a basis, if a replacement pipe is proposed where the existing X42 pipe is to be replaced by a section of X60 pipe (SMYS 60,000 psi), what will be the minimum wall thickness for the new pipe?

In the above situation, all other design parameters are unchanged. Hence, there are two options to arrive at the same result.

The first method is to make the calculation using the formula used earlier (Eq. (2.2)) to determine the pipe wall thickness for a 60,000 psi pipe.

$$PD/2sFET = t$$

Substituting the values in the formula given above we have

$$1480 \times 30/(2 \times 60,000) \times 0.72 \times 1 \times 1) = t$$
$$t = 0.5138 \text{ in.}$$

The other simpler and quick method is to find the ratio of the existing pipe's SMYS and the proposed pipe's SMYS. See the following example:

Ratio of two SMYS = 42,000/60,000 = 0.7.

Now multiply this ratio (0.7) with the original X42 pipe wall thickness (0.734):

0.7 × 0.734 = 0.5138 in.

A 0.5138-in. pipe wall of X60 pipe will meet the required MAOP criteria.

Note that this method works only if all other parameters are held constant.

With substitution, there are various other factors that need to be considered—if there is heavy transition between two pipe walls resulting in welding Hi-Lo issues and if the substituted pipe grade requires a new set of welding procedure and qualified welders.

2.7 EXPANSION AND FLEXIBILITY

The expansion and flexibility of the pipeline, especially if constructed above ground, are of importance for designing a pipeline. If the temperature of the environment is significantly lower than normal, this will affect the thermal expansion of pipes in arctic regions. However, if the pipe is heated and insulated, there might be cause to consider thermal expansion during design.

Thermal expansion may not be the major concern but flexibility certainly is, in arctic pipeline. The thawing of the permafrost can cause the soil to give way, so removing the support for the pipeline. The nominal use of flexibility control practices such as bends, loops, and offset methods is not adequate for arctic conditions; additional methods are used, and these are discussed further in this book.

The flexibility of pipe is calculated for the pipeline as a whole and is based on the modulus of elasticity, E_c, at the given temperature of the pipeline. The combined stress is calculated as follows:

$$S_E = (S_b^2 + 4\,S_t^2)^{0.5} \qquad (2.3)$$

where

S_E = combined expansion stress (psi)
S_b = resultant bending stress (psi)
S_t = torsional stress (psi).

The resultant bending moments (M_b) and torsional bending moments (M_t) used for calculations are in pound-inch (lb-in.).

The stress intensity factors and section modulus of the pipe (in.^{3}) are obtained from tables, and they vary for fittings of various types of transitions in weldings and reducers used in the pipeline system.

Normally, the combined expansion stress range S_E given in Eq. (2.3) is limited to less than 0.72 times the SMYS of the pipe. Other factors that limit this are the design factor and the temperature rating discussed above. The external loads in addition to the weight of the pipe itself are the wind load and the weight of the content of the pipe. The sum of all these should not exceed 0.72 times the SMYS. The maximum reaction is often written as the R' of the pipeline after cold springing has been determined from the flexibility calculations. The variable required to determine the reaction is spring factor C_s, modulus of elasticity E_c, and reaction R' corresponding to the full expansion range based on the modulus of elasticity.

In general, the value of R' is taken as 0.6 if the value of C_s is between 0.6 and 1.0.

The relationship is as follows:

$$R' = (1 - 0.666\ C_s)R$$

2.8 UNSUPPORTED SPAN OF PIPE

When discussing arctic pipes, we pointed out that a large section of pipes are often laid above ground and at a significant height to pass through marshy land, or to provide natural crossing for wild animals. The question that arises in such a case concerns the length of pipe that can be spanned between two supports. This span length (L) can be calculated.

The pipeline engineer should be able to calculate this span using a simple formula, as discussed here (Eq. (2.4)).

The principle is to equate the maximum bending moment caused by the loading to the resisting moment of the pipe.

The operative equation is

$$L = (SI/Wc)^{0.5} \qquad (2.4)$$

where

L = length of the span (ft)
c = radius of the pipe OD (in.)
I = moment of inertia of the pipe (in.^{4})
S = maximum allowable fiber stress of the material (lb/in.^{3})
w = uniform loading per unit length of pipe (lb/ft).

The loading on the pipe includes the sum of pipe weight, fluid weight, dead weight, and wind load.

In the following hypothetical example, we will calculate the maximum allowable span for a 24-in. diameter pipe that has an estimated loading of 45 lb/ft. The pipe has a maximum allowable fiber stress of 30,000 lb/in.^{2}.

From the above we have

c = the radius of pipe = 12 in.
$I = 28.14$ in.^{4}
$w = 45$ lb/ft
$S = 30,000$ lb/in.^{2}

Substituting in these values we get

$$L = \{(30,000 \times 28.14)/(45 \times 12)\}^{0.5}$$
$$= (844,200/540)^{0.5}$$
$$= (1563.333)^{0.5}$$
$$= 39.539 \text{ ft}$$

A span of a maximum of 39.5 ft is allowed; however, it will be prudent to span the pipe no longer than 39 ft.

There are several nomographs that are available and used for an easy and quick calculation; however, this method is most handy if the engineer is not shy of numbers.

2.9 BUOYANCY OF PIPES

The weight-to-volume ratio of pipes is often very low. This makes the pipe floatable on water. This is further complicated if the fluid the pipe carries is of low density like gas. This causes a pipe to float in marshy and watery areas across which it might run. This is especially of concern in arctic regions where wintery, frozen land becomes a mushy mass of marshland when the ambient temperature rises. When this happens, the support to the pipe is lost, and additional stresses are developed in the pipe and on weld joints, ultimately leading to pipe failure.

This demands that the buoyancy of pipe is determined and that suitable corrections are provided. Often, this is done by providing extra weight on the pipe to induce negative buoyancy.

The buoyancy of a pipe is calculated as follows:

$$B = (D/3)\ (D - 32t) + 11t^2 \qquad (2.5)$$

where

B = buoyancy (lb/ft)
D = outside diameter of the pipe (in.)
t = wall thickness of the pipe (in.)
W_c = density of the concrete (lb/ft^3)
t_1 = thickness of the concrete coating (in.)
W_m = density of the mud (lb/ft^3).

●●●───────────────────────────────────────

Example

In the following example, we will show the calculation of 100 lb/ft negative buoyancy for a 30-in. diameter pipe that has a 0.375 in. thick wall. The pipe passes through mud that has a density of 63 lb/ft^3 and the density of concrete is 142.86 lb/ft^3.

Buoyancy of pipe

$$
\begin{aligned}
B &= 30/3\ \{30 - (32 \times 0.375) + 11\ (0.375)^2 \\
&= 10\ \{30 - 12\} + 11 \times 0.1406 \\
&= 180 + 1.546875 \\
&= 181.546875\ \text{lb/ft (positive buoyancy)}
\end{aligned}
$$

To achieve 100 lb/ft negative buoyancy, this pipe needs to be weighed down with concrete coating. To determine the thickness of the required coating, we calculate further.

The following equation is used:

$$B = 10.7\{(DW_m/2000) - t\} + t_1 D\{(W_m - W_c)/48\} \qquad (2.6)$$

Substituting in the known values we find the following:

$$
\begin{aligned}
B &= 10.7\{(30 \times 63/2000) - 0.375\} + t_1 \times 30\{(63 - 142.86/48)\} \\
100 &= 10.7 \times \{(1890/2000) - 0.375\} + t_1 \times 30\{-79.86/48\} \\
49.9125t_1 &= (10.7 \times -0.374.055)/100 \\
t_1 &= 0.103259/100 \\
&= 2.068\ \text{in.}
\end{aligned}
$$

Thus in this case, a concrete coating of 2.068 in. thickness will give the pipe 100 lb/ft negative buoyancy.

The concrete coating is permeable and can make pipe handling very difficult. Alternative options are available for holding the pipe down in place. One such is the use of specially designed sacks that are filled with preformulated gravel and sand. These sacks are paired, slung over the pipe, and strapped on in order to provide the necessary weight. The calculations of required weight are the same as for the above. The calculation can be done to determine the weight and the required number of sacks which may be used to control the buoyancy of the pipe.

Other methods include use of anchors. All these options have their advantages and disadvantages. The selection of methods is based on the specific project need.

The additional weight that is required is calculated to keep the pipe from being buoyant. Often this involves determining the density of the water in which the pipe is laid and the buoyancy of the pipe therein. Then, additional weight is added, and the two sets of data are correlated to determine the required weight to control the buoyancy of the pipeline. The engineers and designers must identify the locations where controlling pipe buoyancy is required and provide the necesssary means of control of the pipe sections in question.

2.10 CALCULATING THE STRESSES IN THE PIPE AT THE TRANSITION FROM BELOWGROUND TO ABOVEGROUND

A pipeline belowground is considered fully restrained, unless it faces some process upheaval, while a pipe aboveground is unrestrained. The stresses and deflections occur at the transition point.

The following values have been assumed for the discussion:

- Young's modulus, $E = 29 \times 10^6$ psi
- Poisson's ratio, $\nu = 0.3$
- Coefficient of thermal expansion $= 6.5 \times 10^{-6}$ in./in.°F.

The analysis of the stress and deflection transition locations resulting from the internal pressure and temperature changes helps determine anchoring requirements. This is a significant step in designing a pipeline. Analysis of the longitudinal deflections helps determine whether or not an anchor block is required. The forces required to be imparted by an anchor block are determined. The purpose of an anchor block is to maintain the pipe in a fully restrained condition.

In the subsequent paragraphs, we shall discuss with the help of some examples the calculations for an anchor block.

2.11 FIRST SCENARIO, REDUCING STRESSES WITHOUT ANCHOR BLOCK

We have a pipe that is buried and coming aboveground at point X. In this case, there is a length of pipe (L) between point X and point Z (fully unrestrained), immediately before the pipe emerges from the ground, which is in transition from being fully restrained to unrestrained. Then, there is a pipe section from point Z aboveground; the point being fully unrestrained.

The section of pipe is fully restrained and is subjected to pressure and temperature changes and therefore suffers tensile stress due to the Poisson effect as indicated by the following relationship:

$$\sigma_{PO} = \nu \sigma_H \tag{2.7}$$

where hoop stress

$$\sigma_H = PDi/2t$$

$P =$ pressure in psi
$Di =$ internal diameter of pipe and
$t =$ pipe wall thickness
and a compressive stress due to a temperature change of

$$\sigma_{TE} = E\alpha\Delta T$$

where

$\sigma_{TE} =$ Compressive stress affected by temperature,
$E =$ Young's Modulus I steel (lb/in.^{2})
$\Delta T =$ the change in the temperature.

The net longitudinal stress therefore at point X will be

$$\sigma_{LA} = \nu \sigma_H - E\alpha\Delta T \text{ psi}$$

and the strain at point $X\varepsilon_A$, being fully restrained, will be zero:

$$\varepsilon_A = 0$$

In the unrestrained section of pipe at point Z, the longitudinal stress (σ_{LB}) caused by only internal pressure will be half the hoop stress.

$$\sigma_{LB} = \frac{\sigma_H}{2} \text{ psi}$$

The associated strain will be a combination of the following:

- Temperature effect

$$\varepsilon_{TE} = \alpha \Delta T$$

- Pressure effect

$$\varepsilon_{PR} = \sigma_{LB}/E = \sigma_H/2E$$

- Poisson effect

$$\varepsilon_{PO} = \nu \sigma_H/E$$

The net longitudinal strain at point Z will therefore be

$$\varepsilon_B = \alpha \Delta T + \sigma_H/E(0.5 - \nu) \text{ in./in.}$$

The transition of stress and strain between points X and Z is assumed to vary as a linear function of length. To establish the length L over which the transition occurs, the longitudinal resistance of the soil is obtained. It is assumed that any tendency to move will be counteracted by constant and opposite soil force (F_s). An average design value for soil is taken as

$$F_s = 80(D_o/12)^2 \text{ lbf/ft}$$

Between point X and point Z, the equilibrium of forces exists and therefore

$$F_s L = A_m(\sigma_{LB} - \sigma_{LA})$$

Deriving from the above we get

$$L = \{A_m(\sigma_{LB} - \sigma_{LA})\}/F_s$$

Total moment at Z will be an average strain between X and Z over the length of L or

$$\delta = \varepsilon_B/2(12L) \text{ in.}$$

2.12 SECOND SCENARIO, REDUCING STRESSES WITH ANCHOR BLOCK

This is a scenario wherein an anchor is required to contain the longitudinal deflection; the stress will be distributed. The transition from being fully restrained to unrestrained will occur at the anchor instead of over a length L as discussed in the previous scenario. The resultant force on the anchor will be in equilibrium as indicated by the following expression:

$$F = (\sigma_{LB} - \sigma_{LA})A_m \text{ lb}$$

An increase of pipe wall thickness after the anchor block will not affect the force as the decrease in stress will be compensated for by the increase in the area of the metal. It is clear that this force is equal to that which would produce a deflection of 2δ in the equation ($\delta = \varepsilon_B/2$ (12L) in.) discussed above.

Strain-based and limit states design is also used where necessary. Basic principles of strain-based design are introduced here.

2.13 STRAIN-BASED DESIGN

The strain-based design is based on limit state design principles. The safety aspect is established on the basis of the variability associated with strain demand given by design requirements on one side and strain capacity on the other. The strain capacity is in fact the pipeline's intrinsic value.

The concept of strain-based design takes into account the differential ground movement caused by frost heave and thaw settlements in arctic soil. The longitudinal strain combined with plastic circumferential elongation is a very common phenomenon in such load conditions. The longitudinal strains—tensile, compression, or multiaxial—accrue with plastic deformation developing in more than two coordinates within a cross section and are not uniform. This imposes some challenging demands in relation to the pipe material and the design. The loading in strain-based design tends to apply a given displacement rather than force to the pipeline. Most often, the displacement imposes a radius of curvature to the pipeline, a result of which is a bending moment. The Canadian Standard Association (CSA) Standard Z662 in its 2003 edition addresses strain-based design. Primary bases of limit state design and strain-based design principles are introduced.

The design value of load action S_d is compared to the design value of the resistance of material R_d. Thus,

$$S_d \leq R_d$$

S_d is the sum of all individual load contributions multiplied by safety (partial) factors $\gamma S, i \geq 1.0$.

$$S_d = \sum \gamma S, i$$

where S,i is the individual load contributor. The design value of the resistance is defined as the quotient of normal (or characteristic) resistance R and the partial factor $\gamma R \geq 1.0$.

$$R_d = R/\gamma R$$

The partial factors represent the statistical scatter of both action side and resistance side and are often different.

Thus, a limit state design condition can be expressed as follows:

1. Factored (increased) maximum tensile strain demand $\leq$ factored (reduced) tensile strain capacity.
2. Factored (increased) maximum compressive strain demand $\leq$ factored (reduced) compressive strain capacity.

Even if the strain-based design is not used, some primary stress analysis is conducted for the conventional design process. This will be discussed further, in part.

Flexibility and stress analysis are conducted, which may consider criteria to develop such specifications as necessary. Some of the most common specifications that are developed are related to the following specific topics.

- Pipe wall thickness
- Maximum allowable temperature differential in restrained sections of pipeline
- Maximum allowable cold spring reaction on equipment attached to flexible piping
- Allowable free span support for axially restrained sections
- Maximum allowable support for stress design of unrestrained sections
- Minimum required flexibility in partially or fully unrestrained sections.

The stresses are calculated. Hoop stress used in stress analysis is calculated according to the following mathematical relationship:

$$S_h = (PD/2t_n) \times 10^{-3}$$

where

S_h = hoop stress (MPa)
t_n = pipe nominal wall thickness (mm), less corrosion allowances, if any

P = design pressure (MPa)
D = outside diameter of pipe (mm).

The flexibility calculations are based on the modulus of elasticity (E_c) at the lowest expected pipe temperature. For carbon steel and high strength low alloy (HSLA) steels, the following values are used for calculating flexibility.

- The modulus of elasticity (E_c): taken as 207,000 MPa (30×10^6 psi).
- The linear coefficient of expansion (α): $12 \times 10^{-6}/°C$ ($7.3 \times 10^{-6}/°F$).
- Poisson's ratio (ν): taken as 0.3.

The combined hoop (S_h) and longitudinal stresses (S_L) are calculated to ensure the stability of pipeline. Hoop stress due to design pressure, combined with the net longitudinal stress due to the combined effect of pipe temperature changes and internal fluid pressure, is limited, $\leq 0.90 \times$ pipe SMYS and temperature factor (which is practically 1 for the arctic environment). In this calculation, the compressive stresses are taken as negative and tensile stresses are taken as positive in the algebraic equation.

Other stresses to be calculated are as follows:

- Combined stress for restrained span
- Anchor restraints
- Stress for unrestrained portions of pipeline systems
- Load on pipe-supporting elements to design support and braces
- Cover and clearances
- Crossings (rail and road utility crossings).

Special environmental crossings, for example, the wildlife tracks and waterways, are discussed further in this chapter.

The design challenges that face engineers include climatic and environmental conditions. The design engineering group starts with understanding the soil and topography of the land, in the process of identifying the challenges associated with each of them, and detailing the action as to what needs to be done to mitigate those challenges. The success of design and construction effort hinges on these primary and important activities, especially for pipeline in arctic regions. The geological survey team and engineers work in close coordination to identify each patch of land type and associated hazards. This may require review of the data in several published works and trips to the field including locating natural landmarks, natural features of the land, and wildlife crossings. This involves aerial surveys during all seasons, use of all terrain vehicles, boats, sledges, and of course, walking the field. Diligent notes are prepared with corresponding GPS locations and sketches. Photographs are taken for future review during the design phase.

As stated above, the designing phase starts with the survey of the proposed pipeline route, and determining the makeup of the soil on the route. The route of a pipeline should be thoroughly analyzed using data from survey charts, maps, GPS, and aerial pictures, thus enabling potential hazards to be identified.

Similarly, for offshore pipelines, the selected route should avoid existing underwater objects such as sunken ships, underwater faults, slide and ice scoring areas, and environmentally sensitive areas, especially the migratory routes of wildlife, should be identified. For offshore route selection, a field hazard survey should be performed, identifying potential hazards.

Soil sampling and other means are used to determine soil types along the route. Decisions are made based on the location of thaw-stable soils, and the pipeline is buried in the conventional manner. Permafrost presents special problems during construction and pipeline operation. Therefore, the occurrence of permafrost on the right of way must be accurately mapped, and this should be done early enough in the design process to allow for rerouting if that is an option. Since ditching in permafrost is very expensive, any requirements for below-ground construction need to be known as early as possible.

Flow assurance for all stages from startup to shutdown must be considered in pipeline design during the lifetime of the pipeline. This is to determine whether adequate flow can be sustained throughout the design life of the pipeline under all expected flow conditions for the range of pressures, temperatures, fluid phase conditions, and fluid properties. The contingencies for slower output must be considered. As production proceeds over time, oil moves slowly; this increases the risk of separation of water from oil. This water can freeze in the pipeline, increasing corrosion and the possibility of failure of pipe.

The Alaska pipeline, which started production at a level of 2 million barrels a day to 560,000 barrels a day in 2011, faces similar challenges if production falls below 350,000 barrels a day.

2.14 CONSTRUCTION METHODS

Pipelines are laid either aboveground or conventionally belowground, and in frozen belowground areas. In areas of thaw-unstable soils, and where heat from the oil in the pipeline might cause thawing and consequent loss of soil foundation stability, the pipeline is insulated and elevated aboveground by means of a unique support system. Factors that challenge an engineer when designing the pipeline route are numerous; some of them present unique design and construction challenges. These are identified and described here.

2.15 PROTECTION OF TUNDRA

The construction method should protect the environment and create minimum disturbances to it. The arctic environment includes tundra, which is a treeless area between the icecap and the tree line of arctic regions, having permanently frozen subsoil and supporting low-growing vegetation such as lichens, mosses, and stunted shrubs which support various fauna. To protect the tundra, construction is often carried out in winter; temporary snow or ice pads are developed for supporting construction equipment and activities. These pads are designed to prevent any contact of equipment and spoils with tundra. Ditching is done using work pads to minimize contact. Low-bearing and track-mounted equipment is often used to minimize the footprint of the construction activities.

2.16 GRADE AND SLOPES

Special consideration is required where there are significant side slopes in the areas containing soils that may have excessive ice content. These slopes may become unstable upon thawing. This restricts the width of a right-of-way. Extreme caution is advised during construction in these areas. As a precaution, granular backfill can be used to surround the pipe to contain the instability of soil. The backfill will ensure drainage and lower the level of the groundwater table.

2.17 CONSTRUCTION IN WARM PERMAFROST

We have discussed the subject of permafrost in the design section. As we have noted, this poses special challenges during construction. In warm permafrost and other areas where heat might cause undesirable thawing, the pipeline is laid aboveground. The section of pipeline in this environment is designed over vertical supports that are placed in drilled holes or piles driven into the ground. To provide stability to theses piles and supports from thawing permafrost, they are provided with two "heat pipes." These are generally about 2-in. diameter pipes and contain anhydrous ammonia, which vaporizes belowground, rises and condenses aboveground, removing ground heat whenever the ground temperature exceeds the temperature of the air. Heat is transferred through the walls of these "heat pipes" to aluminum radiators fitted on the top the pipes, which cools the air and circulates it.

For the sections of the pipeline route that lie on permafrost, where right of way drainage and erosion control measures discussed above and granular backfill is added, similar measures are planned for other ditching operations. These measures include extra local fill to avoid ditch developing over the pipe in the initial couple of years after construction. The settlement of thawed soil will create a shallow ditch over the buried pipeline, sometimes exposing the pipeline, and disturb the drainage system. To prevent this from happening, the following actions are taken.

• Establish monitoring and maintenance program to ensure remedial action is taken on a priority basis.
• If possible, plant some grass or shrubs to hold the soil and reduce erosion.
• Use ditch plugs along the existing drainage system to ensure rain is directed away from the pipeline.
• Use local material for breaking the slope of the right of way, which will help direct the runoff water away from the pipeline.

The pump stations and compressor stations are constructed over permafrost soil, creating similar problems for the compressors and pump station if soil is allowed to thaw. To prevent such occurrences, the soil is protected from excessive construction disturbances; piping and utilities are installed aboveground. The equipment and structures that may cause thaw of the permafrost are also installed aboveground.

In areas where the ground at burial depth is not permafrost, for example, water-saturated soil near a river, engineers consider insulating the pipe to seal it off from the freezing earth. However, this can result in the development of frost heave, which can damage the pipe.

Insulated belowground pipes may not respond well to conventional cathodic protection measures since the thermal insulation usually is also an excellent electrical insulator which interferes with the protective electrical field. In any event, cold pipes will naturally have a much lower rate of corrosion than warm pipes, so cathodic protection may not be necessary, except if mandated by regulation.

As stated in the introduction to the subject, "conventional belowground" is where the pipeline is laid in trenches dug in the ground with a layer of fine bedding material and covered with prepared gravel padding and soil fill material. These trenches often are up to 4−16 ft deep depending on the dimensions of the pipe and the soil conditions. Cathodic protection anodes of zinc ribbons or magnesium are buried alongside the pipeline to protect it from external corrosion.

In frozen belowground areas, pipeline is laid either aboveground or belowground similar to the case for a conventional pipeline. In areas of thaw-unstable soils, and where heat from the oil in the pipeline might cause thawing and consequent loss of soil foundation stability, the pipeline is insulated and elevated above ground by means of a unique support system. Thus the factors that challenge an engineer designing the pipeline route are numerous.

In those areas that have thaw-unstable soils and which are also on the migratory route of wildlife or which have highways crossing them or present the danger of rock slides and avalanches, pipelines are insulated to protect the permafrost from the heat of the pipeline and buried. There is potential for under-insulation corrosion in such pipes; measures must be put in place for their periodic inspection and monitoring.

In contrast to the above, there are locations that are deliberately frozen by the use of coils containing chilled brine to maintain frozen and stable conditions around the buried pipeline. Special refrigeration plants are designed and installed at these locations.

In flood plains, pipelines require scour protection from floating ice in shallow waters. Various types of mats are used for this purpose.

2.18 DESIGN AND CONSTRUCTION IN THAW SETTLEMENT OF GAS PIPELINE

The thaw settlement problem described earlier is still present, but in a gas pipeline, it can be countered by chilling the gas. Belowground gas pipelines have been successfully operated in permafrost by chilling the gas to less than 0°C to maintain the surrounding soil in a frozen state. This creates another problem: frost heave, in which water migrates toward cold fronts in unfrozen ground, freezes to form ice lenses, which, as more water arrives, grow and lift and bend the pipeline. Ideally, the temperature of the gas and the pipeline wall ought to match the ground temperature exactly, so both frost heave and thaw settlement are avoided, but this is not easy to manage and control.

2.19 DITCHING AND BACKFILLING

In suitable soil, chain trenchers are the most suitable equipment for this purpose. They have the following advantages over traditional equipment such as backhoes and rippers.

1. Chain trenchers are best for cutting into rocks and permafrost.
2. Width of the trench can be controlled to greater accuracy by chain trenchers.
3. Requirement for ditch padding is significantly reduced, and in some cases, it is totally eliminated.

These advantages further aid the reduction in environmental footprint, reduction in backfilling material, and ease of machinery maneuvering close to the pipeline.

In construction of pipeline, other components will be used such as mainline valves, pumping unit and discharge valves, terminal inlet and outlet valves, tank farm valves, scrapper and launchers, and storage tank meters. These components are similarly designed and procured as in a conventional pipeline project except that care is taken to select material suitable for extremely cold and demanding service. Their installation also faces similar challenges of permafrost and thawing. The design engineers must consider these challenges and address them to avoid failure.

2.20 CONTROL VALVES

Generally, the sizing of control valves is based on 20–70% of valve travel for the given design condition. The control valves are designed to maintain throttling control at maximum and minimum flow rates. The throttling valve size is often designed to 50% of the pipe size. The valve actuators are selected with an action from wide open to shutoff without showing excessive oscillation or hunting. Often, forged construction body and bonnet are specified.

2.21 PUMPS FOR CRUDE OIL PIPELINES

Often, the pumps with one standby unit are specified. In case of any failure, the operating load is matched with pump and gear efficiency. In designing the pump, the analysis of the crude oil and the ambient temperature of the site are considered. The shaft output is determined with the matching engine kilowatts power (kW) at a given RPM rating. Gas engines are also used. Often, these pumps are mounted on a structural steel skids ready for installation on site.

When selecting a suitable pump system, care should be taken to ensure that the engine has low emission including low NOx, has low maintenance cost, and is fuel efficient. This is in addition to the engine block casting being of good quality nodular cast-iron and an efficient crank shaft and bearing with reliability at high cylinder pressures. The crank shaft should be able to keep the bearing load to a minimum for efficient working of the pump engine. The efficiency of other parts such as piston, piston ring, the cylinder and its antipolishing ring, and connecting rods must be specified and be suitable for the service demands. The cooling and lubricating system, automation, and control should be specified.

Russia has vast gas reserves in the Yamal peninsula that juts into the Arctic Ocean and has built an extensive network of pipelines to bring gas to the industrial cities of Siberia. There are reports that in European Russia, and, further, into Europe, the early designed Siberian pipeline thawed the very wet ground it was buried in; and as the ice support was lost, the pipeline lost lateral support and buckled dramatically.

In later designs, care was taken to address the thaw conditions. Gulf Interstate Engineering, Moscow division, has contributed in the engineering of several recent pipeline segments and facilities projects in the Yamal peninsula.

Oil and gas development beneath the arctic seas creates a need for underwater pipelines. The thaw settlement and frost heave problems essentially disappear because the upper boundary of the permafrost rapidly drops away from the shore and because the seabed temperature regime is far more stable and little affected by the seasons. The principal problems with arctic underwater pipelines is the ice gouging and construction damages.

2.22 DESIGNING AND CONSTRUCTION TO DETER ICE GOUGING

Ice gouging of the seabed occurs when large ice masses, mostly ridges, formed by collisions between different ice sheets, move over the seabed. Technically, these are not "icebergs"; these ice sheets drift into shallow water, run aground, and are then pushed along by more ice behind them. The pressure of the ice pushing, cuts deep gouges into the seabed; up to 200 ft (about 5 m) deep and over 2000 ft (about 50 m) wide gouges have been reported. A simple calculation will show that the force developed by the push of these massive ice sheets is sufficient to cut into steel pipe wall and cause severe damage. Several cases of such damage have been reported not only in arctic regions but also from the Caspian Sea area where pipeline has been laid in the shallow waters found at the northern edges. The extent of damage caused can be compared with that caused by ship and barge anchors pulling on pipeline resting on the seabed. The damaging force of barge anchor pull is well known. The result is very expensive in terms of both dollars and environmental damage. The force exerted by gouging ice is in the range of about 10 times that of an anchor pull. The engineering design solution to this is to trench the pipe deep enough to allow the ice sheets to passes harmlessly over it, but it has to be safely below the deepest gouge, and somewhat deeper because the seabed soil beneath the ice is heavily deformed. A layer of additional soil placed directly above the pipe adds further safety to the pipeline.

2.23 DESIGN AND CONSTRUCTION FOR SUBSEA ARCTIC PIPELINES

Like all arctic projects, finding a reliable and economical construction method for underwater pipelines is a harder problem. New developments are made; innovations and modifications of existing technologies play a major role in this field. It also depends on location: the Arctic is a huge area, with a widely varying climate, and it makes no more sense to think of a unique "arctic" environment than it would to think of a unique "tropical" environment. The engineer has some freedom to choose the season of the year. In some places, there is a long open-water season, and there a pipeline might be constructed by conventional laybarge, reeling, or tow methods. The difficulty is to be confident that the construction vessel can be brought to the site, can lay the pipeline, and can get out before the ice consolidates, failing which the vessel would be trapped over the following winter. It is not only the trapping of the vessel that is of concern, but when it is retrieved after a long and hard winter freeze, it needs to be thoroughly inspected for damage that the crushing ice might have caused to the hull and body. This is expensive and causes delay and uncertainty in the work schedule.

In other areas, the winter ice may be stable and might have relatively less movement. At such locations, the pipeline can be made up on shore and dragged by an ice-based winch along the seabed under the ice, if necessary into a pre-excavated trench. In 1978, the Drake Gas Field at Melville Island, Canada, used this technique to install a pipeline bundle off the Arctic coast. This project used a plough to excavate the trench, and when many years later the well was plugged and the site was cleaned up, the plough was disposed of by burying it in a hole. After the passage of several thousands of years archaeologists may be puzzled at such a finding and struggle to find the reason for it.

Another alternative is to weld together the pipeline on the sea ice, cut a trench through the ice, and lower the pipeline to the seabed with side booms, similar to conventional land construction. This technique was successfully used for laying pipeline to the Northstar artificial island off the north coast of Alaska.

The bottom pull methods is preferred, particularly if the pipeline is running to an artificial drilling and production island, or to a bottom-founded caisson platform. Conditions where sufficient sea ice is not available in summer to support a laybarge, and also the winter ice is not stable enough for ice-based construction, usually because the site lies close to constantly moving ice pack.

2.24 DESIGN AND CONSTRUCTION OF PIPELINE CROSSINGS

Wildlife migratory route crossings are often elevated sections to a height of about 10 ft aboveground to allow for unfettered migration of animals. For the same reason where the soil supports the weight of the pipeline, it may be buried so that the annual migratory routes of wild-life crossings are unaffected.

Bridges are designed to carry the pipeline over waterways and estuaries; different types of bridges are designed to support pipeline weights and movements. Some of them are classified as follows:

- Orthotropic box girder
- Plate girder
- Suspension
- Tied arch bridges.

Similarly, roads are constructed to reach different sections of the pipeline both during and post construction periods.

The buoyancy of pipeline crossing shallow waters and rivers is calculated and compensated for by adding additional weights; these can be in the form of concrete weights or specially designed sacks filled with measured, predesigned weighted material for keeping the pipeline anchored to the floor. Other pipe anchoring systems such as the helical pile system have also been developed and can be considered on engineering evaluation.

Other important points that need to be considered and are designed along with the pipeline itself may include some permanent facilities such as the following:

- Access roads
- Pressure relief station
- Marine terminal
- Pig launching/receiving facilities
- Pipe shoes
- Pump/compressor stations
- Topping units
- Pipeline valves
- Airfields.

2.25 THERMAL EXPANSION AND CONTRACTION

2.25.1 Addressing Thermal Expansion and Contraction of Pipeline

Linked with the subject of stress calculation is the fact that pipes expand and contract with variations in temperature. These changes build stress in the pipeline. The change in length is 0.798 in./100°F change in temperature. This change can consist of expansion as well as contraction.

The simple calculation of contraction is illustrated in the following example.

●●●——

Example

In an unrestrained pipeline of 1000 ft, a change of temperature occurs from $+32°F$ to $-40°F$ in winter. Estimate the contraction of the pipeline.

Total drop in temperature is $72°F = \{+32 - (-40)\}$

$$\text{Contraction} = 0.798 \text{ (from above)} \times \text{(length of pipe/100)} \times \text{(change in temperature/100)} \tag{2.8}$$

Substituting the values,

$$\begin{aligned} \text{Contraction} &= 0.798 \times (1000/100) \times (72/100) \\ &= 0.798 \times 10 \times 0.72 \\ &= 5.7456 \text{ in.} \\ &= 5.75 \text{ in.} \end{aligned}$$

Due to a change in the temperature of crude oil, the surrounding pipe experiences thermal expansion or contraction. Factors such as tie-in temperature, which is the actual pipe temperature at the time when final welds were made to join strings of pipe into a continuous line; hot position, which is the maximum oil temperature (145°F) in the pipe; and cold position, the pipe at minimum steel temperature ($-60°F$), are considered. The importance of these varying temperatures and resulting expansion and contraction of steel pipe can be illustrated by the following practical observations obtained in the field.

Each 40 ft length of pipe expands up to 0.031 in. with every 10°F rise in temperature and contracts the same distance with each 10°F drop in temperature. Longitudinal expansion of a typical 720 ft straight aboveground segment at minimum tie-in temperature to maximum operating temperature is recorded as being about 9 in.

Since the pipes are anchored at intermittent distances, the longitudinal expansion converts to lateral movement in a zigzag configuration for aboveground sections of pipeline.

The vertical supports on which the pipeline is laid are spaced according the span calculations and provide anchoring support at designed spacing.

The buried pipes also face thermal expansion and contraction conditions but since they are relatively restrained by the soil, and not free to move as a result, they accumulate what is called thermal stress. This is a longitudinal stress, and the value is often taken as 25 ksi. The design considers the pipeline movement that might be caused by an earthquake, in the seismically active areas.

In addition to the thermal expansion and contraction, other aspects of pipe stability are also considered; one of them is the stability with respect to the uplift bucking. Several experts have written papers addressing this important aspect of the design of arctic pipelines. In a frost heaving environment, pipe stability with respect to uplift buckling can be dependent on the vertical pipe displacements induced by frost heave and the stabilizing effect of the soil to pipe interaction. Some experts have developed a simplified yet very conservative criterion for the prediction of upheaval instability that helps to illustrate the issue. Since these criteria are overly conservative, they should be used with full assessment of assumptions and for practical applications.

According to these experts, the driving axial forces causing upward displacement are dependent on internal pressure and the temperature differential in the pipe wall. The analysis incorporates the following three major simplifications.

1. They formulate a differential equation with the assumption that there is no axial movement; the presence of the substantial axial force developed in the pipeline by that movement is ignored.

    The limitations of this simplification are that the movement at a bend is acceptable as long as the movement is constrained by the soil. The axial force has a significant, stabilizing effect on the upheaval buckling behavior. It is initially due to the tangential soil to pipe interface friction or skin friction, caused along the pipe surface as the pipeline is displaced axially toward the potential upheaval location.

    Experience has shown that axial shear forces along the surface of a buried pipeline develop at quite small axial displacements. Additionally, the transverse upheaval displacement is generally large relative to the distance along the pipeline over which this displacement takes place, what would be termed a catenary or cable tension effect. The pipeline is therefore also subjected to large displacement extensions at the upheaval location, which contribute further to the stabilizing axial force developed in the pipeline with progressive upheaval displacement.

2. The second simplified assumption is about the upheaval deflection profile. This is idealized as an arc of a circle with a constant curvature.

    The limitations of this simplified assumption are that the flexural rigidity of the pipeline is ignored in the formulation of the instability criterion.

    The pipe is not bent to a uniform curvature for any significant length along the deflection profile. The varying curvature along the pipe develops significant flexural forces due to the stiffness of the pipe segment, and these forces resist upward displacement.

3. The third assumption is about the instability criterion which is presented in a closed form expression assuming that the pipeline would remain elastic for the upheaval displacement.

    It is well known that displacements of a pipeline subjected to upheaval buckling may be governed by strain criteria and that the axial strains may typically extend into the nonlinear, plastic range. A criterion based on elastic behavior may therefore result in a severely restrictive limitation on the allowable displacement due to upheaval buckling.

The examples below show that these points must be considered for a realistic assessment of the potential for upheaval displacements. Consider the following example and the difference in the results of the two approaches.

An insulated 20-in., 0.375-in. wall is covered with 1 in. thick insulation, installed with a minimum cover depth of 4 ft and backfilled with soil with a unit weight of $0.0637505 \, \text{lb/in.}^3$ ($17.3 \, \text{kN/m}^3$). It was designed for an internal pressure of 1440 psi (9930 kPa). This pipeline has operated safely with a temperature differential of 150°F (65°C) for over 10 years. Conventional construction practices were followed, with field overbends of radius of 40D, i.e., a bend radius of 20.32 m, meeting CSA Z662 requirements. To establish the safe limit for a proposed increase of the operating temperature, overbend stability analyses similar to the numerical analysis described by Nixon and Burgess (1999) were carried out using CSA Z662 recommended limit states factors and strain-based design criteria. The results of the analyses showed that the overbends for this pipeline will remain stable for the combined loading from the specified design pressure and a temperature differential of 172°F (78°C). For comparison, the stability of these overbends was also assessed using the criterion proposed by Palmer and Williams. Substitution of the relevant values for this pipeline showed that upheaval instability is predicted to occur for a curvature of 0.000838 in.$^{-1}$ (0.033 m^{-1}), which corresponds to a bend radius of 1189 in. (30.2 m), or about 59D.

It is noteworthy that the destabilizing forces have very different relative magnitudes for a small, relatively low pressure oil pipeline, and a larger high pressure gas pipeline. For the smaller 12.75 in. (0.324 m) diameter oil line, the axial forces due to temperature differential are significantly greater than the force due to a normal range of internal operating pressures. The reverse is true for a larger gas pipeline operating at higher pressures.

Two additional points should also be considered in assessing the stability of a pipeline with respect to uplift buckling. In general, for cross-country pipelines, the terrain is undulating and pipelines include many overbends and sidebends. A more rigorous analysis of the uplift buckling phenomenon would recognize that many of the overbends and sagbends have the effect of relieving the axial load on the pipe within the virtual anchor length of those bends.

As pointed out in earlier discussions, frost heave and upheaval can in some circumstances interact to threaten the integrity of arctic pipelines. The loss of pipe stability with respect to upheaval is not only an arctic pipeline issue and not caused only by the frost heave; it may be a cause for pipe failure but it may not be the exclusive cause. On the Trans-Alaska Pipeline System (TAPS) fuel gas line, upheaval has taken place on several occasions while the pipeline was in service.

Remediation was successfully carried out by excavating alongside the upheaval, laying the pipeline down in the excavation and covering it with granular and insulating material.

The Norman Wells pipeline has also been successfully remediated by covering the uplifted section with granular material.

Nixon and Burgess in their paper published in 1999 presented detailed results from the well-documented case history of an uplift event. The results contain a rigorous numerical analysis indicating conditions under which differential frost heave might provide the initiating conditions for the loss of pipe stability from uplift buckling.

Such studies and reports are often created using particular parameters, and to address specific points; readers are advised to pay detailed attention to those variables and points and use engineering evaluations to arrive at their own conclusions.

2.26 DESIGN OF COMPRESSORS AND PUMP STATIONS

Compressors and pump stations are designed to make operations safe and efficient. The sizing is done with the intent of making the system efficient throughout the range of operating conditions. The main compressors and pump buildings are located on the operating company's property. Safety of the adjacent properties from fire and noise must be kept in mind while deciding the location of stations and designing the size of the unit.

There should be access to the firefighting equipment and sufficient open space around the buildings. These buildings should be sufficiently ventilated, while appreciating the fact that in cold conditions, exhaust gases or noxious vapors which are flammable and hazardous tend to accumulate and settle low and are not free to escape easily, unless forced air ventilation is provided. Such ventilation becomes a hazard if it is not provided with emergency automatic shutoff. The building exits are designed for easy egress in case of emergencies.

The prime movers, gas compressors, pumps, auxiliaries, accessories, and control systems, are designed for safe operation in the given set of operating conditions. The separators are fitted with a liquid removal system; separators are also installed with an automatic shutdown device in case a slug of liquid enters the compressor system. Often, the alarm system and automatic systems are installed in conjunction. Separators are designed to meet ASME Section VIII requirements with a design factor of 0.5 or less. Compressor station gas piping is of suitable grade of steel and is designed to meet ASME B31.3 requirements. The instrument, control, and sampling piping are often of stainless steel.

2.27 DESIGN OF COMPRESSOR STATIONS

2.27.1 Compressor Building Exits

For safety considerations, two exits are provided for compressor building basements, each operating at the main floor of the compressor building. All elevated walkways or platforms over 10 ft or more aboveground or floor level are provided an exit with fixed ladders.

The maximum distance of an unobstructed escape path from any point on an operating floor to an exit is limited to about 70 ft. Such exits are designed with an unobstructed doorway so located to provide a convenient escape to a place of safety.

The latches to the door are designed to readily open from the inside without a key. Swinging doors located in exterior walls are designed to swing outward. Fully tempered or safety glass is used for exit door windows.

2.28 DESIGNING A GAS COMPRESSOR UNIT

Compressor station prime movers and gas compressors, together with all auxiliary equipment, accessories, and control and support systems, are designed for the safe and efficient operation of the unit throughout the range of operating conditions. Gas compressors are designed for continuous service within a range of operating conditions up to the maximum prime mover output.

2.29 LIQUID-REMOVAL EQUIPMENT

Where the presence of liquids is anticipated, provision is made to protect the compressors against the introduction of such liquids in quantities that could damage them.

Separators are used to protect compressors and are mostly designed to meet ASME's Section VIII Boiler and Pressure Vessels Code Requirements.

They are designed with a manually operable means of removing liquids.

Where there is a possibility that slugs of liquid could be carried into gas compressors, an automatic shutdown device is installed to initiate unit shutdown upon detection of high liquid levels in the station liquid-removal equipment. Design also considers the installation of automatic liquid-removal facilities and a high-level alarm.

2.30 EMERGENCY SHUTDOWN FACILITIES

Compressor stations are designed with emergency shutdown systems that address the following essential requirements.

- The systems are capable of excluding escaped gas from the station and blowing it down the station piping, while continuing to provide a supply of gas necessary to essential station auxiliaries that need to be maintained.
- Blowdown piping is designed to discharge gas in such a way that it will not be any hazard to the compressor station or surrounding area.
- The systems design is capable of providing a means for the shutdown of all gas compressing equipment and gas-fired and electrical facilities in the vicinity of gas headers and compressor buildings. The only exception that could be made is to remain operative where certified electrically safe equipment is installed.

Other code restrictions and conditions are imposed depending on the condition of the location. Engineers are advised to check current code and regulatory requirements of the area they are working in.

Where compressor stations supply gas directly to distribution systems with no other adequate source of gas available for ESD, the emergency shutdown system is so designed not to vent gas. One of the functions of an emergency shutdown system is to prevent unintended disruptions of the gas supply to the gas distribution systems. The design is made to address these issues.

Other objective of the design is to construct a compressor station that is able to prevent the possibility of explosion or fire damage to equipment or services necessary for the satisfactory operation of the station. Designing of protective barriers, providing alternative energy source redundant controls, and remote location of critical equipment are some of the safe design measures.

Compressor stations are provided with means for the automatic control of the gas temperature; via monitoring procedures, if necessary they can shut down if the gas temperature exceeds acceptable safe limits.

Compressor station gas engines that operate with pressure gas injection are designed with automatic shutdown of the engine automatically and cut off the fuel and vents to the engine distribution manifold.

The gas turbines of a compressor station are equipped with the means of automatic shutdown of the unit by the cutting off of the fuel supply.

Fuel gas lines within compressor stations, serving various buildings, including any associated residences, are provided with master shutoff valves located outdoors.

The pressure-control compressor stations' fuel gas system is designed to limit the pressure up to a maximum of 25% above the normal operating pressure of the system, and there is control over this maximum operating pressure.

2.31 STATION PIPING

Compressor station gas piping is designed in accordance with the requirements of ASME B31.3. Emergency valves and controls are to be clearly identified by signs including the functions of aboveground piping such as main service fluids, lubricating oil, water, steam, processing, and hydraulic fluids.

In addition to the above, more safety devices are designed into the system. Some of these include the following:

- Compressor station prime movers other than electrical induction or synchronous motors are required to have automatic devices to shut down the units before the speed of either the prime movers or the driven units exceeds the maximum safe speed established by the respective manufacturers.
- Crankcases of gas engines in compressor stations are required to be equipped with explosion doors or suitable crankcase ventilation.
- Mufflers for gas engines in compressor stations have vent slots or holes in the baffles of each compartment to prevent gas from being trapped in the mufflers.

- Gas compressor units are required to have shutdown or alarm devices that operate in the event of inadequate cooling or lubrication of the units.
- Gas compressor units are required to have devices to prevent the temperature of the discharge gas from exceeding the maximum design temperature of the gas compressor and associated gas piping systems.
- Centrifugal gas compressors that use seal oil as a gas compressor sealing mechanism are required to have an emergency seal oil system sized to permit safe shutdown of the compressor on loss of normal seal oil supply.
- Vibration and surge protection is designed into the centrifugal compressors.
- Thrust loads and moments imposed on mechanical equipment are limited to less than the manufacturer's recommended values or, where such recommended values are not available, the limits specified in API 617 and 618 for centrifugal and reciprocating compressors, respectively.
- Centrifugal and positive displacement gas compressors are designed with an automatic high-pressure unit shutdown or unloading device. The high-pressure sensing point should be located between the compressor and the first block valve on the discharge side of the compressor.

PART 2

Safety and Communication

CHAPTER 3

Safety on Construction Sites

Safety is by far the most important aspect of any work site and more so in the challenging environment of arctic regions. The challenge of working in arctic regions is more serious because of the elements of nature in particular. The working conditions and the terrain are very unforgiving, with temperatures often reaching as low as $-74°F$ ($-58.8°C$), along with the addition of wind chill. The extreme cold limits the ability of human and machine. Movement is impaired and reflectivity is reduced, raising the possibility of mistakes leading to accidents and mishaps which could cause injuries and even death. Cold temperatures, wet weather, and frigid winds can take a drastic toll on the human body, increasing the possibilities of frost bite, hypothermia and trench foot. Occupational Safety and Health Administration (OSHA) provides a cold stress card that describes how to prevent and treat these serious illnesses. The conditions increase the threshold levels for safety hazards (Figure 3.1).

The safety program for a site should include restating of all safety precautions and identifying the possible hazards on the site, which may include a daily walk through the job site and looking out for hazards created by snow and ice. There should be inspection of walkways, work platforms, scaffolds, stairs and ladders, and removal of icicles, especially when temperatures are beginning to warm. If this is not possible, the cordoning off of the area under the icicles should be done until they are no longer a hazard.

Portable heaters are used to keep people warm in confined spaces like field workshops and offices; however, they can be very dangerous if not used properly. Have heaters inspected prior to their use to make sure they work properly. To prevent fire hazard from these heaters they should be placed on a sturdy, fire-resistant surface ensuring that the hoses are protected from damage and excessive heat. The area where the heater is being used must be properly ventilated to allow excess fumes to escape. Combustible materials should not be stored within same area or at the least they should be kept about $10-15$ ft

Figure 3.1 A typical construction site.

away from heaters. Propane and other flammable gas cylinders must be stored upright and chained.

Most importantly, use common sense when working outdoors during the winter months. Take care of yourself and those around you.

3.1 HAZARDS TO CONSTRUCTION WORKERS

The leading safety hazards on site are falls from height, motor vehicle crashes, excavation accidents, electrocution, machines, and being struck by falling objects.

Falls from height are the leading cause of injury in the construction industry. In the OSHA Handbook (29 CFR), fall protection is needed for areas and activities that include, but are not limited to: ramps, runways, and other walkways; excavations; hoist areas; holes; formwork; leading edge work; unprotected sides and edges; and other walking/working surfaces.

The height limit where fall protection is required is not defined. A fall from any height can be dangerous and may result in injury. Protection is also required when the employee is at risk of falling onto dangerous equipment.

Fall protection can be provided by guardrail systems, safety net systems, personal fall arrest systems, positioning device systems, and warning line systems.

All employees should be trained to understand the proper way to use these systems and to identify hazards. OSHA 29 CFR Subpart M lists details of instructions dealing with fall protection.

Construction and transportation vehicle crashes are common occurrences on any construction sites and are another major safety hazard. It is important to be safety cautious while operating these machines and motor vehicles.

Equipment on the job site must have light and reflectors if intended for night use. The glass in the cab of the equipment must be safety glass. The equipment must be used for their intended task at all times on the job site. Further instruction can be found in 29 CFR Subpart O.

Before any excavation has taken place, the contractor is responsible for obtaining suitable permissions and notification to all applicable parties that excavation work is being performed. During excavation, the contractor is responsible for providing a safe work environment for employees and other personnel using the area. The contractor should comply with all standards set forth in 29 CFR Subpart P.

Access and egress is also an important part of excavation safety. Ramps used by equipment must be designed by a competent person, qualified in structural design.

No person is allowed to cross underneath or stand underneath any loading or digging equipment. Employees are to remain at a safe distance from all equipment while it is operational.

Inspect the equipment before every use. Further instruction for excavation can be found in 29 CFR Subpart P.

3.2 APPLICABLE SAFETY REGULATIONS

In the United States, the OSHA sets and enforces standards concerning workplace safety and health.

In the EEC, the European Union Directives provide guidance and laws to protect against construction site accidents.

CHAPTER 4

Pipeline System Communication

The transportation of oil and gas through pipeline systems is accomplished by adding compression stations approximately every 50–100 miles. In addition, safety shut-in valves are also included at critical places in a pipeline. Pipeline pressures and flows are typically monitored and/or controlled with SCADA systems using remote terminal units (RTUs), programmable logic controllers (PLCs), and by satellite, microwave, or telephone communication links. These controls are subject to interruption from loss of communication and interference by knowledgeable third-party intruders.

Seamless bridging and routing capability to connect different systems and subsystem is the major consideration. Reliable communications links from one end of a pipeline to the other end are vital for effective pipeline monitoring and control. In addition, reliable communications is needed to support devices residing in the pipeline route that diagnose and repair pipeline problems. There are developing technologies to improve and enhance communication systems. Nearly all of these developments are designed to retrofit into existing pipeline infrastructure. Appropriate technology is selected for the implementation of the communication links, the end goal being to set up an intra-pipeline communication system that would connect to a monitoring and control center.

The importance of communication cannot be overemphasized especially when the locations are remote and treacherous as they are in arctic regions and distances are not easy to negotiate at all times. The monitoring of pipeline from a safe centralized location is both safe and practical.

In such a harsh environment, a system of remote supervisory control and data acquisition, often referred as the SCADA system, is usually installed. The system usually consists of the following subsystems:

1. A human to machine interface (HMI); this is an apparatus or device that presents processed data to a human operator, and through this, the operator is able to carry out monitoring and control functions.

2. A supervisory (computer) system, gathering (acquiring) data on the process and sending commands (control) to the process.
3. RTUs connecting to sensors in the process, converting sensor signals to digital data, and sending this digital data to the supervisory system.
4. The PLC system used as field devices, being more economical, versatile, flexible, and configurable than special-purpose RTUs.
5. Communication infrastructure connecting the supervisory system to the RTUs.
6. Data acquisition communication infrastructure connecting the supervisory system to the RTU, PLC, and subsystem computers located at different zones on the pipeline.
7. Various process and analytical instrumentation.

4.1 SYSTEM CONCEPTS

The SCADA system is built on scalable open architecture. Most control actions are performed automatically by RTUs or PLCs. Host control functions are usually restricted to basic overriding or supervisory level intervention. For example, a PLC may control the flow of cooling water through part of an industrial process, but the SCADA system may allow operators to change the set points for the flow and enable alarm conditions, such as loss of flow, leak in the pipeline or any segment of it being monitored, and variations in temperature, to be displayed and recorded. The feedback control loop passes through the RTU or PLC, while the SCADA system monitors the overall performance of the loop.

In the system, the primary function starts with data acquisition at the RTU or PLC level and includes meter readings and equipment status reports that are communicated to SCADA as required. Data is then compiled and formatted in such a way that a control room operator using the HMI can make supervisory decisions to adjust or override normal RTU (PLC) controls. Data may also be collected and shared with other data users or stored as historical records through a database management system.

SCADA systems typically implement a distributed database, commonly referred to as a tag database, which contains data elements called tags or points. A point represents a single input or output value monitored or controlled by the system. Points can be either "hard" or "soft." A hard point represents an actual input or output connected to field I/O, while a soft point results from logic or math operations applied to other points. Most implementations conceptually remove the distinction by making every property a "soft" point expression, which may, in the simplest case, equal a single hard point. Points are normally stored as value–time stamp pairs: a value and the time stamp indicate the time when the data value was either recorded or calculated. A series of value–time stamp pairs gives the history of that point. It is also common to store additional metadata with tags, such as the path to a field device or PLC register, design time comments, and alarm information.

4.2 THE HUMAN–MACHINE INTERFACE

The HMI is the apparatus which presents processed data to the human operator and through which the human operator controls the process.

An HMI is linked to the SCADA database and software programs, to provide trending, diagnostic data, and management information such as scheduled maintenance procedures, logistic information, detailed schematics for a particular sensor or machine, and expert-system troubleshooting guides.

The HMI system usually presents the information to the operating personnel graphically, in the form of a mimic diagram. This means that the operator can see a schematic representation of the plant being controlled. For example, a picture of a pump connected to a pipe can show the operator that the pump is running and how much fluid it is pumping through the pipe at the moment. The operator can then change the setting of the pump as required. If a leak occurs, the HMI software will show the flow rate of the fluid in the pipe decreasing in real time. Mimic diagrams may consist of line graphics and schematic symbols to represent process elements or digital photographs of the process equipment overlain with animated symbols.

Various manufacturers and suppliers of SCADA system have developed their specialized designs and they market their systems on that strength. However, the overall and primary system principle remains the same. These manufacturers and suppliers also publish their own details to educate prospective buyers. When tasked with selection of a specific system one should evaluate the project needs against the systems being offered by vendors. The following and most of the descriptions herein are generic in nature.

The Graphical User Interface (GUI) development program used by system integrator or system maintenance personnel to change the way these points are represented in the interface is often the part of the HMI package for the SCADA system. This is presented in various ways and could be an on-screen light representing an actual light in the field, or some complex system which could include a multi-projector display, representing an overview of entire pipeline monitoring stations, giving a very detailed and comprehensive view.

An alarm system is also included in a SCADA system. Often this is treated as most important part of the system.

The system is designed to monitor conditions that are serious enough such that they require attending to without delay. It also records the event that initially triggered the alarm system. The systematology involves an alarm event being identified and consequent action options being initiated. These actions could include sending an email or a text message to let the SCADA operator take cognizance of the event. The alarm event settings also create a situation that compels the operator to take action to deactivate the alarm, and thus the operator cannot ignore the alarm as an action is mandated.

Simple electronic gates can be set in logic (Yes/No gates) form to activate the alarm system. The digital status points are either an alarm condition or no alarm condition (Yes/No logic). These are calculated by a formula based on the values in other analog and digital points. A more upgraded version could include a range of setting, which will be set within a set of data to distinguish between high and low ranges of alarm conditions. The manifestation of the alarm system could be in any form; it could be a siren, a pop-up box on the operator's screen, or a colored or flashing area on a screen. Whatever the method used, the main goal of the system is to draw the attention of operator to that part of the system, so that appropriate action can be taken. The induction of an alarm system should be carefully planned, in particular to avoid a situation where a cascade of alarm events might occur in a short time. If such is the situation, the very object of including an alarm to warn of emergencies will fail. Unfortunately, when treated as a noun, the word "alarm" is used rather loosely in the industry; thus, depending on context it might mean an alarm point, an alarm indicator, or an alarm event.

However, SCADA networks generally have little protection from the rising danger of communication attacks. Hybrid-based cryptography, which is a combination of symmetric AES and asymmetric RSA, is the solution to enabling confidentiality and authentication being placed at each end of SCADA communication. This carries out real-time experimental analysis and results using SCADA network facilities to investigate the timing performance and propagation delays associated with encryption/decryption operation of packets from the master terminal unit (MTU) to RTUs and/or RTUs to MTU.

4.3 HARDWARE SOLUTIONS

As with the system, the hardware selection must be based on the project's specific needs. The parameters must be developed including, for example, the ability to function properly in the arctic environment. The list may include factors like the availability of a power source and the possibility of being cut off for a length of time without maintenance, and still being able to function notwithstanding.

The choice of hardware for a specific task is an important step in getting the full advantages of a designed communication system. SCADA solutions contain distributed control system (DCS) components. The industry phrases, like "smart" RTU or PLCs, are often used to describe the capacity of system that are capable of autonomous execution of simple logic processes. The term smart indicates that they are able to process and execute without involving the master computer. Application of such RTUs and PLC has increased in the industry. The programming and control languages are standardized; a suite of five programming languages, including Function Block, Ladder, Structured Text, Sequence Function Charts, and Instruction List, is frequently used to create programs which run on these RTUs and PLCs. Unlike a procedural language, such as the C programming language or FORTRAN, the IEC 61131-3 requires minimal training due to the resemblance to historic physical control arrays. The development in these areas of improved communication and supervisory controls has enabled SCADA system engineers to improve upon the design and implementation of a program to be executed on an RTU or a PLC. A programmable automation controller (PAC) is a compact controller that combines the features and capabilities of a PC-based control with those of a typical PLC. PACs are deployed in SCADA systems to provide RTU and PLC functions. In many electrical substation SCADA applications, "distributed RTUs" use information processors or station computers to communicate with digital protective relays, PACs, and other devices for I/O, and communicate with the SCADA master in lieu of a traditional RTU.

Since the late 1990s, virtually all major PLC manufacturers have offered integrated HMI/SCADA systems, many of them using open and non-proprietary communications protocols. Numerous specialized third-party HMI/SCADA packages, offering built-in compatibility with most major PLCs, have also entered the market, allowing mechanical engineers, corrosion engineers, electrical engineers, and technicians to configure HMIs themselves, without the need for a custom-made program written by a software developer.

4.4 REMOTE TERMINAL UNIT

The RTU connects to physical equipment. Typically, an RTU converts the electrical signals from the equipment to digital values such as the open/closed status from a switch or a valve, or measurements such as pressure, flow, voltage, or current. By converting and sending these electrical signals out to equipment, the RTU can control equipment, such as opening or closing a switch or a valve, or setting the speed of a pump. It can also control the flow of a liquid.

4.5 SUPERVISORY STATION

The term supervisory station refers to the servers and software responsible for communicating with the field equipment (RTUs, PLCs, etc.), and then to the HMI software running on workstations in the control room, or elsewhere. In smaller SCADA systems, the master station may be composed of a single PC. In larger SCADA systems, the master station may include multiple servers, distributed software applications, and disaster recovery sites. To increase the integrity of the system the multiple servers will often be configured in a dual-redundant or hot-standby form providing continuous control and monitoring in the event of a server failure.

For some installations, the costs that would result from the control system failing are extremely high. Possibly even lives could be lost. Hardware for some SCADA systems is rigged in design to withstand temperature, vibration, and voltage extremes. In the most critical installations, reliability is enhanced by having redundant hardware and communications channels, up to the point of having multiple fully equipped control centers. A failing part can be quickly identified and its functionality automatically taken over by backup hardware. A failed part can often be replaced without interrupting the process. The reliability of such systems can be calculated statistically and is stated as the mean time to failure (MTF), which is a variant of mean time between failures (MTBF). The calculated MTF of such high reliability systems can be on the order of centuries.

4.6 COMMUNICATION INFRASTRUCTURE AND METHODS

SCADA systems have traditionally used combinations of radio and direct wired connections, although SONET/SDH is also frequently used for large pipeline systems. The remote management or monitoring function of a SCADA system is often referred to as telemetry. Some users want SCADA data to travel over their preestablished corporate networks or to share the network with other applications.

Many operators have connected SCADA system to Internet and existing cellular network to monitor their infrastructure along with Internet portals for end-user data delivery and modification. Cellular network data is encrypted before transmission over the Internet.

With increasing security demands there is increasing use of satellite-based communication. This has the key advantages that the infrastructure can be self-contained, not using circuits from the public telephone system, can have built-in encryption, and can be engineered to the availability and reliability required by the SCADA system operator. Modern carrier-class systems provide the quality of service required for SCADA.

RTUs and other automatic controller devices were developed before the advent of industry wide standards for interoperability. The result is that developers and their management created a multitude of control protocols. Among the larger vendors, there was also the incentive to create their own protocol to "lock in" their customer base.

The security of some SCADA-based systems has come into question as they are seen as potentially vulnerable to cyber attacks.

In particular, security researchers are concerned about the following:

- The lack of concern about security and authentication in the design, deployment, and operation of some existing SCADA networks.
- The belief that SCADA systems have the benefit of security through obscurity by the use of specialized protocols and proprietary interfaces.
- The belief that SCADA networks are secure because they are physically secured.
- The belief that SCADA networks are secure because they are disconnected from the Internet.

SCADA systems are essential for the effective functioning of transmission and communication system, and are not immune to threats to their integrity and effective performance. Various users have identified and reported threats that can be grouped under two main points as follows:

1. The compromise of access to the control system through an unauthorized access could lead to the loss of security of data and communication. This access could be through a person or through the control software. The loss of security can be attributed to the changes induced intentionally or accidentally by virus infections and other software threats that may be coming to the host machine.
2. The threat could be linked to the packet access to the network segments hosting the SCADA devices. In many cases, these are very basic and loss of security on the actual packet control protocol can occur. Any source that can generate and send packets to the SCADA device can take control. The real threat is associated with the fact that often SCADA users assume that a virtual private network (VPN) provides the required protection. This assumption of false security is the real threat. It may be noted that physical access to SCADA-related network jacks and switches allows saboteurs to bypass all security on the control software and take full control of the SCADA network. A system of endpoint-to-endpoint authentication and authorization should

be established to protect the integrity and the security of the system. This can be achieved through the use of installed secure socket layer (SSL) or cryptographic technologies.

Many vendors of SCADA and control products have begun to address the risks posed by unauthorized access by developing lines of specialized industrial firewall and VPN solutions for TCP/IP-based SCADA networks as well as external SCADA monitoring and recording equipment. In recent years, the International Society of Automation (ISA) has started formalizing SCADA security requirements by evaluating the requirements, measurements, and other features required to assure security resilience and performance of industrial automation and control systems devices.

4.7 SCADA SYSTEM FOR TYPICAL GAS PIPELINE

In practical application, the SCADA system for a typical gas pipeline is organized in the following way (Figure 4.1).

The pipeline communication and control system is designed to operate from a centralized location through the SCADA system from a master control center (MCC) or a secondary control center (SCC). These centers are designed to remotely monitor and control all pipeline facilities along the pipeline, including compressor stations, gas off-take stations, metering and regulating (M&R) stations, actuated main line valve (MLV), electric utility stations, and other facilities along the pipeline. The SCADA system is also used to connect and monitor a safety instrumented system (SIS).

As stated earlier, the monitoring station for the SCADA system is found in a centralized location, often the pipeline company's Head Offices. The pipeline facilities will be locally monitored and controlled from local control systems (LCSs) or RTUs or local PLCs on specialized packaged equipment. Often the booster compressor stations, gas off-take stations, metering and regulating stations, and main line valve stations are not manned.

A well-designed system should include redundancy. The communication system is designed with redundancy, and is comprised of dual-redundant modules or subsystems, such as servers, networks,

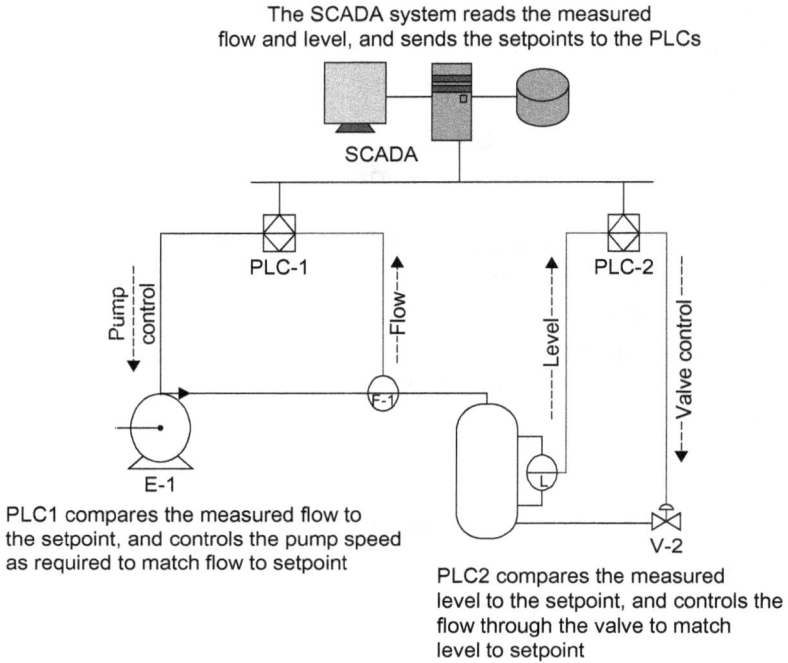

Figure 4.1 Schematic arrangement of SCADA system.

workstations, communication equipment, printers, storage devices, and RTUs. Individual network cards are installed in these devices to connect to each local area network (LAN).

Switchover between redundant equipment or subsystems in case the primary fails is seamless for both the process and the operators, apart from the event/alarm announcing the switchover.

The system is designed to constantly check the statuses of both the primary and the secondary modules and indicate failures of either of them. The system indicates the current primary/backup/start-up/failed status of each component.

4.8 SCADA OPERATOR WORKSTATIONS

Workstations at the MCC and SCC are each designed to include one PC with a number of display screens. The operator workstations provided for the SCADA system support the operator interface and real-time and archival system functions for the SCADA system.

Operator workstations are used to access the process information database of the SCADA, historical data, and data relating to pipeline application software (PAS), and RTUs. Each operator workstation is connected both to SCADA LANs and to the user interface to allow the user to interact with the information in the SCADA system. A typical operator workstation for a gas pipeline is the main interface for all SCADA system functions and may include the following:

- Graphical displays showing the process conditions of the pipeline
- Trends of selected process variables
- Alarm and event management including alarm acknowledgment
- Process trends and analysis displays
- Commands and controls to change the operating state of the pipeline facilities such as opening or closing of valves
- Summaries and reports
- System maintenance and configuration changes
- Programming and system-level access to the servers
- Intelligent cause and effect displays for the station logic with "what-if" analysis
- Gas management system (GMS) displays
- PAS displays
- SIS displays
- Displays related to corporate geographical information system
- Safe start up and shut down guidelines
- Diagnostics of the system up to card level and instrumentation including system-malfunction indications
- Calibration and tuning displays
- Asset manager
- Report view and print
- Communication error displays.

4.9 SCADA SERVERS

SCADA servers are provided in the MCC, SCC, and supervisory station. The MCC and SCC may consist of the following servers:

- SCADA database servers, redundant for the MCC
- Historian database server
- PAS server
- GMS server

- Video wall server
- Web server
- Printer server.

The HCC supervisor station may have the following servers:

- SCADA database server
- GMS server
- Printer server.

4.10 SELECTION OF SOFTWARE

The selection of software is done after the review of the system. The typical key points that may be used to evaluate software for a SCADA system are as follows:

- Software which has been previously installed and provided satisfactory operation and control of similar control systems on currently operational units for at least 2 years.
- Software that requires minimal tailoring with specified parameters to suit the requirements.
- Software for which standard data and performance documentation are available.
- Software for which correction/updating facilities are available to all users.
- Configuration function blocks and ladder logic is in compliance with IEC definitions.
- Any known bugs, faults, or problems to be identified are known before making decisions.

4.11 COMMUNICATION INFRASTRUCTURE SECURITY

The open architecture and high dependence relating to modern SCADA automation also increases security concerns on the use of SCADA systems. The security concerns should be discussed and identified regarding the proposed employment of a SCADA system. The following measures are the common steps to be considered.

- Multiple levels of password protection
- Locking unused ports on network switch
- Firewall at different levels and zones in the SCADA architecture

- Specialized security software
- Disabling unused services on operation system
- Troubleshooting over the network without physical access to the equipment.

4.12 TYPICAL CODES AND STANDARDS

Typical codes and standards applicable to design a typical communication (SCADA) system relate to various industry bodies including the following. These documents must be consulted in detail to ensure that design and effective communication systems are specific to the requirements of a pipeline system.

4.12.1 API
- API 1113   Developing a pipeline supervisory control center
- API 1130   Computational pipeline monitoring

4.12.2 ANSI
- ANSI/NFPA 75   Standard for the protection of electronic computer data processing equipment

4.12.3 NEMA
- NEMA ICS 1   General standards for industrial control and systems
- NEMA ICS 2-230   Components for solid-state logic systems
- NEMA ICS 3-304   Programmable controllers
- NEMA ICS-6   Enclosures for industrial controls and systems

4.12.4 IEC
- IEC 61131   Programmable controllers—ALL PARTS
- IEC 60870-5-101   Telecontrol equipment and systems—Part 5-101: Transmission protocols—Companion standard for basic telecontrol tasks

4.12.5 ISA
- ISA S5.1   Instrumentation symbols and identification
- ISA S5.2   Binary logic diagrams for process operation
- ISA-5.3-1983   Graphic symbols for distributed control/shared display instrumentation, logic, and computer systems
- ISA S5.4   Instrument loop diagrams

- ISA-5.5-1985   Graphic symbols for process displays
- ISA S18.1   Alarm annunciation sequence and specification
- ISA-50.02   Fieldbus standard for use in industrial control systems—Part 2: Physical layer specification and service definition
- ISA RP55.1   Hardware testing of digital process computers
- ISA-RP60.3-1985   Human engineering for control centers
- ISA S71.04   Environmental conditions for process measurement and control systems: Airborne contaminants
- ISA S72.1   LAN industrial data highway
- EWICS-1998   Guidelines for the use of PLCs in safety-related systems
- EWICS-1997   Guidelines on achieving safety in distributed systems

CHAPTER 5

Electrical Equipments

5.1 INTRODUCTION

The design of electrical equipment for arctic pipeline must ensure the temperature compatibility of the machine, equipment, and supporting parts. Common engineering challenges are overheating and providing for additional cooling. While this may be a consideration, given that in summer temperatures do rise high on the scale, a new added challenge is to ensure that very low temperatures in winter do not cause equipment to freeze up or make the material or insulation brittle and so compromise the performance of the equipment. So in designing for arctic conditions, challenges exist in coping with both unusual and normal weather conditions.

We will note that the general design calculations and parameters are same as those used for warmer temperatures except that we have to be seriously mindful of facts associated with extremely low temperature, and material and equipment must be rated for low temperature service and protected from the extreme conditions that might arise.

There are several standardizing bodies that have issued specifications and standards which affect the quality and performance of electrical equipment. Some of these are listed below and must be referenced for the most recent and correct engineering design.

• National Electrical Manufacturers Association (NEMA)
• Underwriters' Laboratories (UL)
• Canadian Standards Association (CSA) (ACNOR)
• International Electrotechnical Commission (IEC)
• Commission Électrotechnique Internationale (CEI)
• Japanese International Standard (JEC)
• Institute of Electrical and Electronics Engineers (IEEE)

5.2 DESIGN OF ELECTRICAL EQUIPMENT

Electrical equipment of various types is used in a pipeline system. Such equipment is used for corrosion control and monitoring; running a SCADA system is discussed in Chapter 4. Electric motors are primary electrical equipment that are installed to run most systems. There are various applications of electric motors, for example, motors for continuous running of fans and blowers, and motors that run the controlled compressors and pumps that operate on an on-off basis. This is called the duty cycle of the motor.

Therefore, choosing an appropriate electric motor also depends on whether the load is steady, varies, follows a repetitive cycle of variation, or has pulsating torque or shocks.

Starting and running torque are the first parameters to consider when sizing electric motors. Starting torque requirements for electric motors can vary from a small percentage of full load to a value of several times the full-load torque. Starting torque varies because of a change in load conditions or the mechanical nature of the machine in which the electric motor is installed. The latter large percentage scenario could be caused by the lubricant, wear of moving parts, or other reasons.

Electric motors feature torque supplied to the driven machine, which must be more than that required from start to full speed. The greater the electric motor's reserve torque, the more rapid the acceleration.

The drive system of an electric motor that uses gear reducers has parts that rotate at different speeds. To calculate acceleration torque required for these electric motors, rotating components must be reduced to a common base. The part inertias are usually converted to their equivalent value at the drive shaft. Equivalent inertia $W_2 K_2^2$ of the load only is found from:

$$W_2 K_2^2 = (W_1 K_1^2) * (N_1/N_2)^2 \qquad (5.1)$$

where

$W_1 K_1^2 =$ load inertia (lb/ft^2)
$N_1 =$ load speed (rpm)
$N_2 =$ electric motor speed (rpm).

Electric motors with straight-line motion are often connected to rotating driving units by rack-and-pinion, cable, or cam mechanisms.

For these electric motors, the load inertia is calculated from the following relations of the required load speed (rpm) and motor speed (rpm).

$$WK^2 = W * (S/2\pi N)^2 \qquad (5.2)$$

where:

W = load weight
S = translation speed (fpm)
N = rotational speed (rpm).

For electric motors in common use, the demand on current is almost directly proportional to developed torque. This relation of current demand to the developed torque does not apply to motors that are squirrel-cage or motors that are used exclusively for acceleration, where high torque is a special requirement.

The relation of current demand to the developed torque is extended to the fact that at constant speed, torque is proportional to horsepower. If load demand is for acceleration or an overload is expected, then the electric motors would have considerable droop, and to meet this type of load factor a matching horsepower is selected. For sizing a motor for specific requirement its performance curve should be evaluated for its starting torque. The staring torque should be able to overcome machine static friction and have sufficient power to accelerate the load to reach full running speed. It should also be able to handle maximum anticipated overload.

The selection of motors is based on the following factors:

1. Determination of power supply
   Power supply includes determination of voltage, frequency, and phase. NEMA classifies nominal power supply voltages in three groups of 240, 480, and 600 V, and the corresponding nameplate markings on motors are 230, 460, and 575 V.
   The frequency of motors is rated to 200 hp or less and a variation is permitted to the rated frequency. Reference to NEMA specifications should be made for more details on the applicable ratings and variations permitted to allowed frequencies.
   Most industrial applications are three phase; however, some single-phase equipment is used.
2. Horse Power (HP) and duty requirements
   The motors can be designed on the basis of the time they will have to perform at their full capacity duty cycle.

Duty cycle is a fixed repetitive load pattern over a given period of time which is expressed as the ratio of on-time to cycle period. When the operating cycle is such that electric motors operate at idle or a reduced load for more than 25% of the time, the duty cycle becomes a factor in sizing electric motors. Also, the energy required to start electric motors (i.e., accelerating the inertia of the electric motor as well as the driven load) is much higher than for steady-state operation, so frequent starting could overheat the electric motor.

3. Speed

The speed of the motor is also an important variable to consider when selecting a motor. Single speed is common; however, variable speed with 2, 3, or 4 speed motors is often selected.

4. Service factors

Signifies the use of the motor over its rating and linked to the motor's torque.

The service factor (SF) is a measure of periodically overload capacity at which a motor can operate without overload or damage. The NEMA standard SF for totally enclosed motors is 1.0.

A motor operating continuously at an SF greater than 1 will have a reduced life expectancy compared to operating at its rated nameplate horsepower.

SF is calculated using its horsepower.

●●●────────────────────────────────────

Example: Service Factor

A 1-hp motor with an SF = 1.15 can operate at 1 hp $\times$ 1.15 = 1.15 hp without overheating or otherwise damaging the motor if rated voltage and frequency are supplied to the motor. The insulation life and bearings life are reduced by the increased SF load.

───

5. Torque

For most electric motors (except squirrel-cage electric motors during acceleration and plugging) current is almost directly proportional to developed torque. At constant speed, torque is proportional to horsepower.

For accelerating loads and overloads on electric motors that have considerable droop, equivalent horsepower is used as the load factor. The next step in sizing the electric motor is to examine the electric motor's performance curves to see if the electric motor has enough starting torque to overcome machine static friction, to accelerate the load to full running speed, and to handle maximum overload.

6. Enclosures

The enclosures of electrical motors are standardized by NEMA; these are listed below:

Drip proof

Designed for reasonably dry, clean, and well-ventilated (usually indoors) areas. Ventilation openings in shield and/or frame prevent drops of liquid from falling into the motor within up to a 15° angle from vertical.

Outdoors installation requires the motor to be protected with a cover that does not restrict the flow of air to the motor.

Totally Enclosed Air Over (TEAO)

Designed with dust-tight fan and blower motors for shaft mounted fans or belt-driven fans. The motor is mounted within the airflow of the fan.

Totally Enclosed Non-Ventilated (TENV)

Designed with no ventilation openings and enclosed to prevent free exchange of air; however, these are not airtight. These motors have no external cooling fan and rely on convection cooling.

The design is suitable where the motor is exposed to dirt or dampness. However, these are not suited to a very moist humid or hazardous (explosive) air.

Totally Enclosed Fan Cooled (TEFC)

Similar to TENV discussed earlier but has an external fan as an integral part of the motor. The fan provides cooling by blowing air on the outside of the motor.

Totally Enclosed, Hostile, and Severe Environment

These covers are designed for use in extreme conditions, where moisture and chemical environments or both conditions prevail. However these are not designed for hazardous locations.

Totally Enclosed Blower Cooled

Designed similar to the TEFC and with external fan on a power supply which is independent of the inverter output. The design gives full cooling even at lower motor speeds.

Explosion Proof Motors

Designed where the ambient motor ambient will not exceed +40 °C. Motors are approved for the classes described below:

CLASS I (Gases, Vapors)

• Group A—Acetylene
• Group B—Butadiene, ethylene oxide, hydrogen, propylene oxide
• Group C—Acetaldehyde, cyclopropane, diethyl ether, ethylene, isoprene

- Group D—Acetone, acrylonitrite, ammonia, benzene, butane, ethylene dichloride, gasoline, hexane, methane, methanol, naphtha, propane, propylene, styrene, toluene, vinyl acetate, vinyl chloride, xylem
  CLASS II (Combustible Dusts)
- Group E—Aluminum, magnesium, and other metal dusts with similar characteristics
- Group F—Carbon black, coke, or coal dust
- Group G—Flour, starch, or grain dust
7. End shield
   The end covers or end shields are three types; they have bolt holes for connecting the cover to the motor body.
   - Type C covers provide a male rabbet and taper holes.
   - Type D covers come with a male rabbet and holes for flange type through-hole connection.
   - Type P covers have a female rabbet and through bolts for mounting the flange connections.

5.3 CONDUCTOR SIZE

The size of conductor used for electrical connections is of importance, this determines the possible resistance as well as drops in voltage across the length of the conductor. The voltage drop is determined for various sizes of copper wire on the basis of its length, the amperage, and a factor that is often viable from standard tables.

The selection of proper size of electrical conductor depends on the load to be carried, the mechanical strength required, and the economics. Once these are considered, the most economical conductor will be the one that equate the energy cost to the cost of copper.

The calculation of the correct size of conductor to use can be made by use of Eqn. 5.3 below, taking into account current in amperes, cost of electrical power, cost of the copper, hours in service, and amortization insurance and taxes.

The standard formula for this calculation is as follows:

$$A = (59) * (I) \{(C_e \times t)/(C_c \times F)\}^{0.5} \qquad (5.3)$$

where

A = conductor cross sectional area (mils)

Mil = A unit of length equal to one thousandth (10^{-3}) of an inch (0.0254 millimeter), used, for example, to specify the diameter of wire or the thickness of materials sold in sheets.

I = current (A)

C_e = cost of electrical power (per kWh)

C_c = cost of copper (per lb)

t = hours of service per year

F = factor for fixed charges.

Take the example of year-long operation at 12 h a day where 100 A is carried at a cost of \$1.01 per kWh and fixed charges are 25% with 2.5% insurance and amortization cost. The cost of copper is \$0.60 per lb.

Using Eq. (5.3):

$$A = (59) * (I)\{(C_e \times t)/(C_c \times F)\}^{0.5}$$
$$= (59)(100)\{(1.01 \times 365 \times 12)/(0.60 \times 0.25)\}^{0.5}$$
$$= 5900\{(4423.8)/(0.15)\}^{0.5}$$
$$= 5900(171.7323)$$
$$= 1,013220.8 \text{ mils (cross-sectional area)}$$

Selection of Materials

CHAPTER 6

Introduction to the Material Selection

For arctic pipelines and facilities with few exceptions, the components used including the pipeline are required to meet the same industry specifications as for normal pipe. The only difference is the additional tests and quality norms that are imposed to meet the challenges of the harsh environment and the serious consequences of failures.

In a pipeline project, various components are used, including linepipe, valves, induction bends, fittings, various forgings like flanges, weldolets, anchor forgings, and gaskets. These items are subject to challenges from environmental and mechanical conditions imposed on them by the design and service conditions. These conditions develop stresses that limit their performance.

In this part of the book, we will discuss the components and parts that are commonly suitable for use in an arctic environment. We will discuss the materials used in a pipeline projects and their inherent properties and how they are affected by the environment in which they are to perform.

The process of selecting material for a pipeline in any environment has two clear paths; these paths are not mutually exclusive but are considered hand in hand often an engineering compromise is made in balancing the two objectives:

1. The type and shape of parts.
2. The physical and mechanical properties of the material these parts are made from.

In the larger scale of the picture, a pipeline system will have the pipe, valves, pumps for liquid transportation and compressors for gas transportation, flanges, fittings, and fabricated equipment like vessels, tanks, boilers, heat exchangers, and many more. Some of these parts may be made of fiber-glass-reinforced plastics, plastics; sometimes nonferrous materials are used but predominantly various grades of steel are employed. These general material types may have different compositions and may be produced by different production processes. Steel, which constitutes the bulk of material used in pipeline system, is available in various grades and types. Selection is made from various grades of steel that are produced in different ways; the production process of each grade or group of material varies significantly. For example, pipe in most cases is produced by rolled plates welded, or may be drawn and extruded out of steel stock to produce seamless pipes.

A valve's body may be made of either of cast steel or forged steel. Flanges with rare exceptions are almost always made of forged steel.

Fittings are often stamped out of wrought steel or in some cases forged steel; they may be seamless or welded construction.

Pumps casings are made of cast steel or forged steel; however, they are (like valves) an assembly of various grades of steel and alloy steels produced from cast, forged, and wrought steel material which are further processed by machining and other secondary processing processes.

In all the above cases, the primary processes may be further supported by one or a combination of secondary processes of rolling, stamping, machining, and welding. The sequence of these operations, the limitations and advantages of each process, and knowledge of secondary processes, help in the understanding of the material.

The determining factors in the selection of materials for a pipeline system are the environment where the candidate material will be put into service, its design life, and the cost.

In the first category—the environment—the factors to consider are the pressure and temperature of the system, and corrosiveness of the fluid carried by the pipeline and components. The exposure implies the conditions that the system will encounter in order to enable it to function for optimum results.

The mechanical properties of a material consist of tensile strength, yield strength, ductility, and toughness as determined by impact tests and supporting tests like hardness and microstructure evaluation. In case the material is subject to cyclic stress, additional toughness tests, and determination of KIC, critical CTOD and critical J values of material and welds may also be specified.

CHAPTER 7

Material Properties in Low Temperature Environment

The challenge of the design and construction of pipeline in any conditions mostly revolves around the safety of men and material. This concern becomes more prominent due to the extreme challenges of the weather in the Arctic, which affects the performance of men and materials. It is the performance of material that is of the focus of this chapter.

The materials used for tools, machines, and pipelines has to face the elements of extreme arctic weather. Their performance is seriously compromised and additional tests are conducted to ensure the safe operation of equipment of all types.

In the following sections we will look at the properties of metals in relation to their exposure to low temperatures.

7.1 DUCTILITY AND BEHAVIOR OF STEEL IN A LOW TEMPERATURE ENVIRONMENT

The selection of material for any specific environment is directly dependent on the material's properties, especially those properties that are affected by that special environment. Metal properties are classified in terms of mechanical, physical, and chemical properties. These are further subdivided in structure sensitive and structure insensitive properties. These properties are described in Table 7.1.

In this discussion, we are concerned only with the structure sensitive mechanical properties of metal.

Metals are favored as a construction material because they offer a combination of mechanical properties that are unique and not found among non-metals. Metals are generally strong, and many of them can be loaded or stressed to very high levels before breaking. One property of metals that is of interest is their capacity to exhibit a high degree of elastic behavior in the early portion of their load carrying capacity. This is a very important property for effective use of a metal as a

Table 7.1 Material Properties		
Metal Properties	**Structure Insensitive Properties**	**Structure Sensitive Properties**
Mechanical	Elastic moduli	Tensile strength, yield strength, tensile elongation, true breaking strength, elastic limit, proportional limit, creep-rupture strength, creep strength, strain hardening rate, tensile reduction of area, fracture strength, fatigue strength, impact strength, hardness, damping capacity
Physical	Thermal expansion, thermal conductivity, melting point, specific heat, emissivity, vapor pressure, density, thermal evaporation, thermoelectric properties, thermionic emission	Magnetic properties, electrical properties
Chemical	Electrochemical potential, oxidation resistance, catalytic effects	

construction material. When metals are loaded beyond their elastic range they exhibit another set of important properties called ductility and toughness. It is these properties and how they are affected by changes in temperature that constitute the core of our discussion.

The specific metal that we will discuss is carbon steel and low alloy steel. It may be noted here that bulk of the material that is used in pipeline engineering derives from this generic group; it is the ductility and toughness of these metals and how they are affected by the variation of temperature that is the subject of this discussion. When we talk of arctic conditions the main thing that comes to mind is the bitter cold and freezing conditions all around. Thus it will not be out of place to assume that the main challenge is the performance of a material in those cold conditions. The emphasis is placed on the behavior of metal under low arctic temperatures. For this purpose, it is essential to know the exact nature of these metal properties and what precisely is connoted by the term low temperature.

Hence we need to be aware of some important terminology applying to material behavior. Some of these terms are now discussed in so far as they relate to fracture mechanics.

First, *ductility* is defined as the amount of plastic deformation that a metal undergoes in resisting fracture under stress. This is a structure sensitive property and chemical composition has significant impact on the ductility of metal.

Second, *toughness* is the ability of the metal to deform plastically, and absorb energy in the process, before fracturing.

7.2 CONCEPT OF TOUGHNESS AND LOSS OF TOUGHNESS IN A LOW TEMPERATURE ENVIRONMENT

Toughness as defined above is the mechanical and structure sensitive property is an indicator of how the given metal would fail at the application of stress beyond the capacity of the metal and whether that failure will be ductile or brittle in nature. Only one assessment of toughness can be made with reasonable accuracy from ordinary tensile testing, and this is to the extent that the metal displays either ductile or brittle behavior. From this it can be assumed that a metal displaying little ductility is not likely to display ductile failure if stressed beyond its limits. The failure in this case would be brittle.

The temperature of metal is found to have a profound influence on its brittle/ductile behavior. The influence of higher temperature on metal behavior is considerable. A rise in temperature is often associated with increased ductility and corresponding lowering of the yield strength. The rupture that takes place at elevated temperatures is often intergranular, and little or no deformation of the fractured surface may have occurred. As temperature is lowered below room temperature, the propensity to brittle fracture increases.

Before we proceed further on the subject, let us take note of some further terminology used in this discussion. ASTM E 616 defines some of the terminology associated with fracture mechanics and testing:

- The term fracture is strictly defined as an irregular surface that forms when metal is broken into separate parts. If the fracture has propagated only part way in the metal and the metal is still intact, cracking is referred to.
- A crack is defined as two coincident free surfaces in a metal that join along a common front called the crack tip, which is usually very sharp.
- The term fracture is used when splitting in metal occurs at a relatively low temperature and metal ductility and toughness performance is chiefly of concern.
- The term rupture is more associated with metal splitting at elevated temperatures.

7.3 FRACTURE TOUGHNESS K_C

As pointed out earlier, basically there are two types of fracture that occur in metals:

1. Ductile fracture
2. Brittle fracture.

These two modes are easily recognized when they occur separately, but fractures in metal often have mixed morphology and this is aptly called mixed mode.

The mechanisms that initiate fracture are shear fracture, cleavage fracture, and intergranular fracture. Only shear mechanisms produce ductile fracture.

It may be noted that, as with the modes discussed earlier, failure mechanisms can also co-occur.

Irrespective of a fracture being ductile or brittle, the fracturing process is viewed as having two principal steps:

1. Crack initiation.
2. Crack propagation.

Knowledge of these two steps is essential as there is a noticeable difference in the amount of energy required to execute them. The relative levels of energy required for initiation and for propagation determine the course of events when a metal is subjected to stress.

There are several aspects to fracture mechanics that tie in with the subject of metal ductility and toughness but this short discussion will not go into detailed information on fracture mechanics, hence these are not discussed in detail; however, some specific related topics are listed below.

- Effects of axiality of stress
- Crack-arrest theory
- Stress intensity representation
- Stress gradient
- Rate of strain
- Effect of cyclic stress
- Fatigue crack

- Crack propagation ($KI_c = \sigma\sqrt{\pi a}$)
- Griffith's theory of fracture mechanics
- Irwin's $K = \sqrt{E \times G}$
- Crack surface displacement mode
- Crack tip opening displacement (CTOD), (BS 5762-1979 and BS 7448 part-I)
- R-curve test methods
- J-integral test method
- Linear-elastic fracture mechanics (LEFM) (ASTM E 399)
- Elastic-plastic fracture mechanics (EPFM)
- Nil ductility temperature (NDT).

The above are commonly taken into consideration when selecting suitable material for pipeline, especially when fatigue and strain are associated with other strains on the pipeline, e.g., sub-sea pipeline and riser technology. In fact some of the specifications (e.g., API 1104 and DNV-OS F101) suggest the use of fracture mechanics to determine the failure behavior of metal in these areas of usage.

Lowering the temperature of metal profoundly affects fracture behavior, particularly of metals that have body centered cubic (BCC) structure. Strength, ductility, toughness, and other properties are changed in all metals when they are exposed to temperature near absolute zero. The properties of metals at very low temperatures are of more than of casual interest, because pipeline, welded pressure equipment, and vessels are expected to operate satisfactorily at levels well below room temperature. The pipeline in arctic regions, buried in permafrost or exposed to elements above ground are one such example; other examples are the imposition of moderate sub-zero temperatures on equipment for de-waxing petroleum, for storage of nitrogen, liquefied fuel gases, and pipelines. Much lower temperatures are involved in cryogenic services where metal temperature falls to $-100°C$ ($-150°F$) and below. The cryogenic service may involve storage of liquefied industrial gases like oxygen and nitrogen. Down near the very bottom of the temperature scale, there is a real challenge for metals that are used in the construction of equipment for producing and containing liquid hydrogen and liquid helium, because these elements in liquefied form are increasingly important in new technologies. Helium in liquefied form is only slightly above absolute zero, which is 1 K ($-273.16°C$ or $-459.69°F$).

Absolute zero (1 K) is the theoretical temperature at which matter has no kinetic energy and atoms no longer exhibit motion. Man has yet to cool any material to absolute zero, so it is not known how metals would behave when cooled to this boundary condition. However, metal components have been brought to temperatures very close to absolute zero. Special challenges are presented to engineers as regards metals and welded components as they would be required to serve in extremely low temperatures.

On cooling below room temperature a metal will reach a temperature where the kinetic energy will be reduced to nil. The atoms of the element will move closer and the lattice parameters will become smaller. All these changes affect the mechanical properties of the metal.

With the earlier discussed aspects of the physics of metals in mind, let us review the behavior of an un-notched specimen without flaws. It may be pointed out that in real life, there is no material without flaw, and that every material has some flaw in it; hence the assumption that a material is flawless is can only be hypothetical in nature. It is the flaw that has to be considered as the initiator of the behavior of a material in a given environment.

The sub-ambient temperature dependence of yield strength σ_o (Rp0.2) and ultimate tensile strength σ_u in a BCC metal is shown in Figure 7.1. The material is ductile until a very low temperature, point A, where yield strength (Y.S.) equals the Ultimate Tensile Strength (UTS) of the material ($\sigma_o = \sigma_u$). Point A represents the NDT temperature for a flaw-free material. The curve BCD represents the fracture strength of a specimen containing a small flaw ($a < 0.1$ mm). The temperature corresponding to point C is the highest temperature at which the fracture strength $\sigma_f \approx \sigma_o$. Thus, point C represents the NDT for a specimen with a small flaw.

The presence of a small flaw raises the NDT of steel by about 200°F (110°C). Increasing the flaw size decreases the fracture stress curve, as in curve EF, until with increasing flaw size a limiting curve of fracture stress HJKL is reached. Below the NDT the limiting safe stress is 5000−8000 psi ($\sim 35−55$ MPa).

Above the NDT the stress required for the unstable propagation of a long flaw (JKL) rises sharply with increasing temperature. This is the

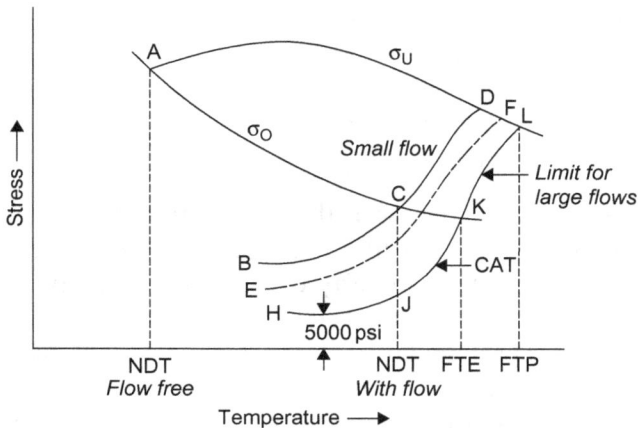

Figure 7.1 Nil ductility temperature.

crack-arrest temperature curve (CAT). The CAT curve defines the highest temperature at which unstable crack propagation can occur at any stress level. Fracturing will not occur at any point to the right of the CAT curve.

The temperature above which elastic stresses cannot propagate a crack is the fracture transition elastic (FTE). The temperature defines the FTE, at the point K, when the CAT curve crosses the yield strength, σ_o curve. The fracture transition plastic (FTP) is the temperature at which the CAT curve crosses the ultimate tensile strength σ_u curve (point L). Above this temperature the material behaves as if it is flaw free, for any crack, no matter how large, cannot propagate as an unstable fracture.

7.4 METAL STRENGTH AT LOW TEMPERATURE

As we have seen as temperature is lowered from room temperature 75°F (24°C or 297 K) to absolute zero, 1 K, the atoms of an element move closer together in accordance with what one would expect bearing in mind the coefficient of thermal expansion. A number of changes occur as a result of the smaller lattice parameters. For example, the elastic modulus increases. In general, the tensile strength and yield strength of all materials increase as the temperature is lowered to the extent that at NDT the yield and tensile strength are equal ($\sigma_o = \sigma_u$). The change in these properties is variable for different metals but change does occur.

When the temperature of low carbon or low alloy steel is lowered the corresponding increase in strength of metals is attributed to an increase in resistance to plastic flow. Since plastic flow is strongly dependent upon the nature of the crystalline structure, it would be logical to assume that metals with the same kind of structure would react in a similar manner.

Toughness tends to decrease as the temperature is lowered, especially for BCC-structured material like steel. Testing is often carried out to measure and monitor this property of steel. The most frequently used test specimen is of the type notched-bar impact, despite the shortcomings of this test. The popularity of the impact test is due to its long established position in standards testing and relatively easy procedure available to laboratories to test in relation to standard Charpy V-notch.

7.5 TYPES OF IMPACT TESTS

Recognition of the influence of temperature, strain rate, and distribution of stresses on the toughness of metals has led to development of several test methods. The initial testing methods were carried out at room temperature; however, currently the lowering of temperature is possible by the use of special baths.

Most of the later developments are named after the originators of the individual test methods, such as Charpy, Izod, and Mesnager. The Charpy and Izod methods are industrially more acceptable tests, and they use notched specimens.

The Izod method uses notched specimens of round or square cross section. The specimen is held as a cantilever beam in the gripping anvil of a pendulum machine. The specimen is broken by a single overload of the swinging pendulum. A stop pointer moved by the pendulum records the energy absorbed. An Izod specimen can be an un-notched bar or it may have a 45° V-Notch in the face struck by the pendulum.

The Charpy design offers a choice as to the type of notches used. There are three options of notch design to choose from, 45° V-Notch, Labeled keyhole, or U-Notch. By far the most common notch type used for steel testing is the 45° V-notch. The specimen is cooled to the required temperature and placed as a simple beam in the horizontal position on the anvil and is centrally struck by the edge affixed to the swinging pendulum, which strikes the specimen on the opposite face of the notch. A single overload breaks the specimen and the absorbed energy is recorded on the stop pointer.

7.6 ENERGY ABSORPTION IN IMPACT TESTING

Energy absorption is a different means to evaluate toughness and gains unanimity in decisions regarding the acceptance of the values obtained. The results are analyzed in a number of ways. The minimum energy absorption is often specified, but it must be noted that typical values differ significantly. In the 1970s, in a project involving hundreds of steel plates, while analyzing the material received for Cv performance it was noted that steel with impact values up to 14 J (10 ft lbf) were quite susceptible to brittle fracture. It was also noted that the plates that exceeded a Cv of 28 J (20 ft lbf) had sufficient toughness to arrest a running crack with a velocity exceeding 6 m/s. These observations were tested in independent laboratories and the results were confirmed. Independent of the earlier work, similar data has been found to be in use in other parts of the world, including in API 5L linepipe specification, for deciding the requirements for low temperature service. The specifications have often specified 20 J (15 ft lbf) as minimum acceptable values irrespective of the design temperature. For example, if a vessel is designed for −20°F and another to −40°F, in both situations the Cv of the selected material selected should have a minimum of 20 J (15 ft lbf). When "leak-before-break" is the main criterion of a design that in most cases includes the pipeline design, the Cv must be set to a higher value at the given design temperature. Generally, the test temperature is also changed to about 5°C below the least anticipated service temperature (LAST), and this practice is common in the design of deep-sea risers and other structures.

A cautionary note: Several ASME/ASTM material specifications do not relate to low carbon steel but to low alloy steels including ASTM A 333 grades 1, 3, 4, 6, 9, and 10 pipes, ASTM A 350 grades LF1, LF2, LF3, etc., and these specify a minimum energy absorption value of 13 J (9.58 ft lbf); this is also true of welding consumables.

7.7 TRANSITION TEMPERATURE FOR ENERGY ABSORPTION

The transition temperature provides somewhat similar criteria for analyzing Cv test results. This method requires Cv test over a range of temperatures from a relatively high temperature where the metal exhibits its best toughness down to a low temperature at which cleavage can initiate. The obtained energy levels are then plotted against temperature. Metal with BCC crystalline structures undergo a precipitous drop in energy over a relatively narrow mid-range span of temperature. The drop in energy coincides with the occurrence of cleavage during the fracturing process.

7.8 TRANSITION TEMPERATURE FOR LATERAL EXPANSION

The extent of plastic deformation that occurs in the Cv specimen's cross section during testing also is a quantifiable value, and this feature undergoes a marked transition in BCC metals with the lowering of the test temperature. When a Cv specimen is broken, a small amount of lateral contraction ordinarily occurs across the width, close to and parallel with the root of the notch; conversely expansion should occur across the width opposite the notch. Changes in dimensions from the original width of a specimen are easily measured, and they are indicators of ductility in the presence of a notch. The extent of lateral expansion opposite the notch is the value presently favored for appraising the capacity of metal flowing plastically during fracture under impact loading.

7.9 DROP-WEIGHT TEAR TEST (DWTT)

Drop-weight tear testing is another way of determining the fracture behavior of steel. The method owes its origins to Naval Research Laboratories tests for determining the NDT of steel in the 1950s.

The process involves use of a 4 in. wide full section steel plate with a sharp notch in the middle. The notch is made with a sharp chisel that has a small ground cutting tip. The DWTT specimen is broken via three-point impact loading using a drop-weight tip or pendulum hammer that has a velocity of no less than 5 m/s (16 ft/s). The specimen is cooled in a bath to the required temperature. The specimen is broken in a single blow.

In this method, no attempt is made to measure the absorbed energy. Evaluation of broken surface is based on the texture of the fractured ends. The final appraisal is based on the percentage of ductile shear facture.

7.10 SELECTING MATERIAL FROM SPECIFICATION AND CODE BOOKS

The pipeline material is selected on the basis of API 5L specification and according to the design specification of, for example, ASME B 31.8 for gas transmission pipelines or ASME B 31.4 for liquid hydrocarbon and other liquids.

For the selection of pipe and related material, the guidance given in API 5L/ISO 3183 is followed for arctic conditions that focus on PSL 2 pipes with CVN tested, and for pipe body ductile fracture propagation resistant properties of steel. A combination of sufficient shear fracture area and sufficient CVN absorbed energy is required for the pipe material to ensure the avoidance of brittle fracture propagation and the control of ductile fracture propagation in pipelines, especially gas pipelines.

Engineering design should take all necessary precautions to ensure that the operating parameters are comparable to the test conditions.

For this purpose, the procurement process should be efficient and must include a full sized specimen and the average absorbed energy level acceptable at given temperature. Commonly the CNV shear percentage shall be greater than 85% and absorbed energy is determined based on the NDT concept discussed earlier. Several acceptable methods are used as specified in Annex G of ASME B 31.8 or as given in ASME B 31.4 section 401.3 and 423.2.3 relating to steel properties.

In Annex G of API 5L there are five different approaches given for determining the minimum acceptable absorbed energy at a specific temperature. ASME B 31.8 also uses four different approaches to establish the absorbed energy level for steel at a given temperature level. The API 5L methods are listed below. The calculation methods and related tables are part of Annex G in API 5L.

1. European Pipeline Research Group (EPRG) guidelines for fracture arrest of gas transmission pipelines
2. Battelle simplified equation

3. Battelle two curve method
4. AISI approach
5. Full-scale burst testing

Similarly, ASME B 31.8 has the following four approaches:

1. Battelle Columbus Laboratories
2. American Iron and Steel Institute (AISI)
3. British Gas Council (BGC)
4. British Steel Corporation (BSC)

It may be noted that there is a significant difference in the number arrived at by different calculation approaches so it is important for engineers to establish the basis of the project requirements and that engineering calculations are based on this.

In addition to API 5L, there are other classes of material that are used in pipeline services. The steel selected for these components also needs to meet the above specifications. It may be noted that recognition of various material specifications and their advantages and limitations in relation to a given environment must be considered in selecting a suitable material.

There are several ASME/ASTM specifications specially tailored for low temperature services, and it is important to check if the specified test temperatures for the metal in use are in agreement with the design temperature of the system. ASTM A/ASME SA105 is not a specification for low temperature material; however, it may be used if all other factors meet requirements and an additional impact test on the material is carried out at a temperature that is in tally with the design temperature. Similarly, ASTM A 106 pipes (grade A, B, or C) must be checked regarding test temperatures because ASTM A 106 is specified for "High Temperature" material and, rightfully, impact testing is not even included in non-mandatory requirements; the same applies for ASTM A 105 forged material discussed earlier. For ASTM A 333 grades 1, 3, 4, 6, 9, and 10 pipes regarding acceptable impact values and test temperatures, the specification must be referenced before arbitrarily using them for any service temperature range. ASTM A 350 LF1 ($-20°F$), LF2 ($-50°F$), LF 3 ($-150°F$) are suitable for low temperature service to the limits set by the specification, but one should check to ensure that the specified energy absorption Cv is in tally with the system design parameters. The point is that an informed selection has to be made. There are several types of boiler quality plate materials specified by the ASTM specifications and ASME codes but not all are suitable for low temperature service, some being metallurgically so designed that they are not suitable for low temperature service; plate material conforming to ASTM A 515 is one such example. Most metals that are fit for low temperature usage are generally tested to 32°F (0°C) unless it is specified otherwise. So the general assumption that all ASME materials are acceptable for service at temperatures up to −20°F will not be correct, unless testing is carried out and the material test report so declares.

API mandates that PSL2 pipes be tested at 32°F (0°C) or any lower temperature as agreed between the buyer and the manufacturer, and are expected to have 20 ft lbf (27 J) absorbed energy. Hence, it is important to determine what the actual test temperature was; the responsibility lies with the engineers to ensure that the test temperature is in tally with the design temperature of the system.

A question is often raised in relation to designing a buried pipeline as to whether one needs to consider low temperature or not. The answer to this question does not concern metallurgy since it is not related to the material properties so much as it is to geography and environment, i.e., the design conditions. Generally, a buried pipeline will not be subjected to very low temperatures unless buried in permafrost, or if there is another cause for low temperature it needs to be determined to what extent during the life of service. If the temperature is ever in the critical low range, it will be prudent to identify those conditions and take them into account while selecting the pipeline material.

Similar considerations apply to above ground pipe and components; above ground valves flanges and pipes etc. are more exposed to the weather; and they also carry similar products so they have a greater propensity to be subjected to low temperature during their lifetime. The following must be taken into account:

- If they are insulated
- If they are heated
- If there is any possibility of depressurization that would lead to extensive temperature reduction.

A multiplicity of factors affect the understanding of material behavior in extreme stress conditions, and all possible must be identified and addressed.

Based on primary information about what constitutes the basics of material selection discussed in this chapter, we will proceed to discuss the details of various pipeline materials in the remaining chapters.

CHAPTER 8

Line Pipes

In Section 2.1, we introduced the concept of strain-based design. This concept also covers the process of line pipe material selection, if the design is based on the strain-based concept. At this point, it is necessary to know the strain capacity of the material, which is expressed through the material's stress—strain curve and its uniform elongation properties. In this application, the concept of strain-based design is not only associated with the longitudinal stress, but the transverse stress—strain behavior of the material is also considered. The effect of temperature on aging of the pipe is taken into account based on the Arrhenius equation. The concept using strain-based design in selecting a line pipe material is not limited to axial strain but includes (circumferential elongation) transverse strain—stress behavior. This is important as the pipe is constantly aging and the yield to tensile ratio changes as well. An adequate ratio and sufficient uniform elongation in the stress—strain curve of the pipe material is necessary. This is especially important when higher grades of pipe materials are considered.

The Arrhenius equation gives "the dependence of the rate constant k of chemical reactions on the temperature T [in absolute temperature kelvins] and activation energy E_a"; energy/mole are the units.

$$k = A\, e^{-E_a/RT} \tag{8.1}$$

where A is the preexponential factor, simply called the prefactor and R is the universal gas constant. Alternatively, the equation can also be expressed as:

$$k = A\, e^{-E_a/k_B T} \tag{8.2}$$

This version of the equation uses energy/molecule as units.

The above introduction to the strain-based design in the selection of material is one way to make proper material selection and it is beyond the requirements of the code and will require mathematical modeling of pipe and weld stress—strain interaction. On normal application, however, the requirements given in codes will meet the design challenges.

In the introduction of this section we have said that in the design and construction of arctic pipeline and facilities — with few exceptions — the material including the pipeline are required to meet same industry specifications as normal pipeline. We will discuss some of these common properties and point out the additional requirements for arctic conditions.

8.1 METALLURGICAL CONSIDERATIONS FOR LINE PIPE STEEL

We have discussed the mechanical properties that are important for pipeline low-temperature service in arctic regions; to achieve those properties, the steel used to make the pipes must conform to some important control processes.

The damage to the environment and the response time required to contain and repair any failure can be immense. This leads to the importance of the material—the steel—that is used to make pipelines. It is important that such steel is made to the most stringent quality standards. A clean steel making process is used, and the consistency of the steels' performance in a given environment is first established. And above all, the steel should have the necessary strength and ductility suitable for the design conditions.

This requirement of arctic pipeline also coincides with the demand to transport crude oil and gas by line pipe at higher operating pressures to increase the capacity. This requires steels that are characterized by a combination of high strength and high toughness.

Increasing the strength of line pipe steels allows taking advantage of strength in design calculations to reduce the wall thickness, which reduces the weight of the steel. To meet these demands, steel industries have developed high-strength microalloyed line pipe steels. Normally, with the increase in higher yield strength, an associated loss in the fracture toughness occurs leading to low formability and the possibility of increased stress-induced cracking. Thus, high strength in conjunction with high toughness and formability are the primary needs of the line pipe.

New alloys have been designed and process improvements made to align alloying with thermomechanical processing. This combination plays an important role in the evolution of the final microstructure and associated mechanical properties. Alloy constituents such as manganese (Mn), niobium (Nb), vanadium (V), titanium (Ti), molybdenum (Mo), nickel

(Ni), chromium (Cr), and copper (Cu) are commonly employed in line pipe steels to obtain the desired microstructure and mechanical properties. This alloying process is very carefully established, and judicious selection of alloying elements is made to obtain the maximum benefit in relation to mechanical properties in tandem with reduced cost of the alloy. In the process employed, the alloying elements are reduced to obtain low carbon equivalent (CE) to ensure good field weldability. On the other hand, alloy additions such as Cr, Cu, and Ni are added to obtain strength in severe corrosive environments.

Controlled thermomechanical processing is considered to be the primary route for the development of API grade steel for line pipe. This process gives desirable and fine-grained microstructure. It allows a combination of high strength and toughness to be achieved by either alloy design or accelerated cooling, or a combination of the two. The ultimate microstructure is governed by processing parameters such as

- Reheating temperature
- Percentage reduction
- Deformation temperature
- Cooling rate
- Coiling temperature.

This is because fine austenite grains, substructure, and dislocations in austenite effectively promote the transformation to fine ferrite. The primary grain refinement in controlled rolling is achieved through the recrystallization of austenite during deformation and the use of microalloying elements, such as Nb, which precipitates in the form of fine carbides and inhibits grain growth.

Thermomechanical processing of line pipe steels can lead to anisotropy in mechanical properties, especially the yield strength and toughness. This is, however, also dependent on the chemical composition, microstructure, and crystallographic texture. These steels tend to develop strong fiber textures during controlled rolling, which involves processes like:

- Deformation
- Recrystallization
- Transformation.

The relationship between microstructure, texture, strength, and toughness is extremely important in spiral welded pipe and has been extensively studied in the steel industry. Research and development laboratories have studied the mechanical properties and the isotropy phenomena of these steels using a combination of electron microscopy and crystallographic texture analysis in a higher range of API grade pipelines steels, which underscores the importance of the relationship between anisotropy in mechanical properties and texture.

The manufacture, testing, and classification of line pipes are predominantly controlled by API 5L, a specification which is now also an ISO specification (ISO 3183). It is strongly advised that the pipeline engineering obtain the latest edition of this specification and use it for design, construction, procurement, and quality control of their projects. The API pipes for pipeline system are primarily classified on the basis of material's yield strength and the process of making the pipe. In general, the processes differ for seamless and welded pipes. The welded pipes are further classified on the basis of the welding process. The other basis of classification for all pipes is the physical attributes such as diameter and wall thickness. Identification is done on the basis of an alphanumeric system (e.g., X42, X56, X70). Some grades like Grades A and B are the exceptions; these do not refer to the yield strength but like all other grades, they too have specified minimum yield strength (SMYS). Grade A steel has a SMYS of 30,000 psi and Grade B has a SMYS of 35,000 psi. These are further classed on the basis of product specification level (PSL); there are two levels of product quality, PSL 1 and PSL 2. Grade A pipes are not included in the PSL 2 category. Pipes produced to meet PSL 2 requirements are different from PSL 1 pipes on the basis of mandatory impact testing. Imposition of upper limits to tensile and yield strength and some welding processes, specifically for example a PSL 2 pipe may not be low-frequency electric resistance welded but must be welded by high frequency welding (HFW) process which is different from electric resistance welding (ERW). The API 5L standard has reclassified the term ERW as EW. The text below explains the difference between LFW and HFW electric welded pipes. Similarly, welding processes like laser and continuous welding are not applicable for manufacturing PSL 2 pipes. There are welding processes that can be used for grades up to X80 but these are not specified for higher grades of line pipes. Since PSL 2 pipes have greater control over chemical analysis and mechanical properties and are mandated to be Charpy impact tested, it is logical that for the harsh environment of the Arctic, PSL 2 pipes will be the preferred choice, and hence we will concentrate on this category of pipe for this discussion.

ERW has, for many years, been used for making longitudinal seam welds in steel line pipe, principally for use in low-grade pipeline applications. The advent of high frequency induction (HFI) techniques has led to significant improvements in weld quality, which in combination with greater control of the raw material's chemical composition has led to the production of high-quality line pipe suitable for more stringent applications in oil and gas pipelines.

Although this product form has been used for many years in difficult service environments, and in the North Sea for unproblematic service applications, there is some resistance to the use of ERW line pipe, particularly in stringent service applications. This lack of confidence in general is based on historical problems such as those relating to the reliability of electric resistance welds, pressure reversals and preferential weld line corrosion, and susceptibility to stress corrosion cracking.

The quality which can be achieved with modern HFW line pipe has improved dramatically with improvements in strip quality and greater understanding of the welding process and nondestructive testing technology. It offers closer dimensional control, which is of great value in pipe laying, and potentially significant cost savings over seamless pipe for similar applications.

The advantage of the HFW process over low-frequency ERW process is that a high frequency weld has a very narrow heat affected zone (HAZ) in relation to the current flow and thus heat is confined to a very small area. In the HFW process, two proximity conductors under a magnetic core are placed opposite each other on the edges of the pipes to be welded. This proximity causes the edges to heat; as they are heated, a force is applied to bring the two faces together and a weld is made. To heat a similar area with conventional ERW (EW in API 5L) would require significantly higher current, and would result in wider HAZ (Figure 8.1).

8.2 SUBMERGED ARC WELDED LINE PIPES

Other process may include the welding with submerged arc welding (SAW) process; welds are made from both the inside as well as the outside, and welding is done from both sides, generating another term, DSAW, which is often used in the industry. Line pipes, especially over a diameter of 24 in., are manufactured using this welding process. Weld

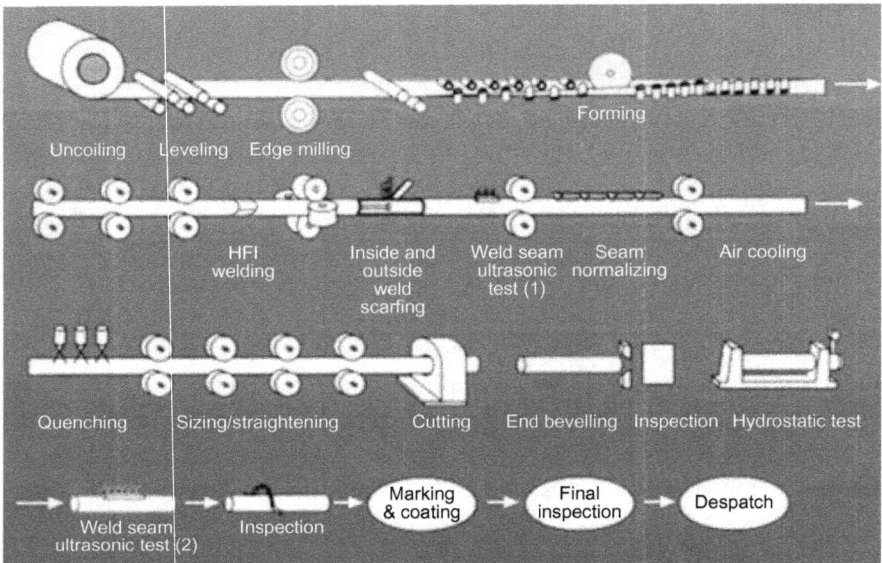

Figure 8.1 Typical HFW (HFI) production flowline.

positioning is also a factor in specifying the type of pipe; for example a straight long weld along the length of the pipe, which is termed longitudinal submerged arc welded (LSAW). If the pipe is made by helically twisting a long steel coil, the weld will run along the abutting faces of the coil forming the pipe; such a weld, often made by use of the SAW process from inside as well as the outside seam, is termed helically welded pipe (HSAW). Both these terms are rechristened in new editions of API 5L as SAWL and SAWH; note the relocation of adjective "helical" and "longitudinal" in the new names. Table 8.1 lists various processes used to produce line pipes.

8.3 CLASSIFICATION OF LINE PIPES

Line pipes are also classified on the basis of their mechanical properties, and this classification is based on the concept of minimum specified yield strength (SMYS), which is often the basis for all engineering design calculations. It may be noted that the SMYS of a grade is a fixed number and is used to define pipe as X42, X56, X70, etc.

Table 8.1 Acceptable Line Pipe Processes

Type of Pipe	Description or Definition
SMLS	Seamless pipe produced by hot forming process, after hot forming, cold sizing, and finishing are carried out
CW	One longitudinal weld produced by continuous welding
LFW	Pipe produced with low-frequency (<70 kHz) EW process
HFW	Pipe produced with high frequency (> 70 kHz) welding process
SAWL	Submerged Arc Welding process—longitudinal seam
SAWH	Submerged Arc Welding process—helical seam
COWL	Longitudinal seam pipes produced by combination of metal arc and SAW process
COWH	Horizontal seam pipes produced by a combination of metal arc and SAW process
Double-seam SAWL	Pipe made in two halves and both longitudinal welds are made by SAW process
Double-seam COWL	Pipe made in two halves and both longitudinal welds are made by combined metal arc and SAW process

Table 8.2 Properties of Steel Line Pipe

Pipe Grades	Pipe Body Seamless or Welded Pipes		SAW, EW, and COW Weld
	Minimum Yield Strength in MPa (psi)	Minimum Tensile Strength in MPa (psi)	Minimum Tensile Strength in MPa (psi)
A25 (L175)	175 (25,400)	310 (45,000)	310 (45,000)
A25P (L175P)	175 (25,400)	310 (45,000)	310 (45,000)
A (L210)	210 (30,500)	335 (48,600)	335 (48,600)
The above grades are available in PSL 1 only			
The grades below are available in both PSL 1 and PSL 2			
B (L 245)	245 (35,500)	415 (60,200)	415 (60,200)
X42 (L290)	290 (42,100)	415 (60,200)	415 (60,200)
X46 (L320)	320 (46,400)	435 (63,100)	435 (63,100)
X52 (L360)	360 (52,200)	460 (66,700)	460 (66,700)
X56 (L390)	390 (56,600)	490 (71,100)	490 (71,100)
X60 (L415)	415 (60,200)	520 (75,400)	520 (75,400)
X65 (L450)	450 (65,300)	535 (77,600)	535 (77,600)
X70 (L485)	485 (70,300)	570 (82,700)	570 (82,700)

Table 8.2 lists the details of various grades of steel according to API 5L/ISO 3183, and the required minimum tensile and yield strength of material and weld.

API 5L gives tolerances on pipe dimensions, for example diameter, the out of roundness, and the wall thickness in tables of various editions. In the 44th edition, these tables include Table-9 (Permissible outside diameter and wall thicknesses), Table-10 (Tolerances for diameter and out of roundness), Table-11 (Tolerances for wall thickness), and Table-12 (Tolerances for random length pipe, etc.). It may be noted that these tolerances are generalized acceptable levels or ranges, and in some specific cases the engineering must consider either accepting the given tolerances or looking elsewhere to suit their specific requirements. This includes changes in the chemical composition of various elements and variations in mechanical properties and dimensional tolerances.

8.4 PSL 1 vs. PSL 2

One of the main differences between PSL 2 and PSL 1 pipes is the mandatory CVN absorbed energy requirements for the pipe body of PSL 2 pipes. Table-8 of API 5L gives the minimum requirements for various grades of PSL 2 pipes from 20 in. (508 mm) diameter to 84 in. diameter. These values are for normal service line pipes. For more stringent requirements, for example for cold temperature service and where ductile fracture conditions are expected, higher values may be calculated and specified.

8.5 DETERMINATION OF PERCENTAGE SHEAR THROUGH DWT TEST FOR PSL 2 WELDED PIPES

The pipe material is expected to be sufficiently ductile to resist brittle fracture and it must be established that such material has the ability to resist fracture propagation in gas pipelines. To establish these properties, the drop weight tear test (DWTT) is carried out. The average of two shear tests is specified to be $\geq 85\%$, at a given test temperature. A good shear value in combination with acceptable CNV values gives confidence in materials' ability to resist fracture.

More details of ductile fracture are discussed in Annex G of API 5L. Readers are advised to refer to current versions of API 5L/ISO 3183 for more updated properties of line pipes.

API 5L has added specific normative annexes to address additional requirements for line pipes. They address the requirements for:

A. Jointers
B. Manufacturing procedure qualification for PSL 2 pipes

C. Treatment of surface imperfections
D. Repair welding procedure
E. Nondestructive inspection for other than sour service or off-shore service
F. Requirements for couplings (PSL1 pipes only)
G. PSL 2 pipe with resistance to ductile fracture propagation
H. PSL 2 pipe ordered for sour service
I. PSL 2 pipe ordered as "Through the Flowline" (TFL) pipe
J. PSL 2 pipe ordered for offshore service
K. Nondestructive inspection for pipe ordered for sour service and/or offshore service.

As their name suggest, these normative annexes are specifically added to address special requirements. There are four more informative annexes detailing information on:

L. Steel designations as used in Europe
M. Correspondence of terminology between ISO 3183 and its source documents
N. Identification and explanation of deviations
O. API monogram use and importance.

8.6 ORDERING A LINE PIPE

Section 7 of API 5L, 44th edition, has an extensive 55 point checklist of information that the purchaser should provide to the supplier. The purchaser can use this list to prepare their project-specific specification and base the purchase order on the information provided.

In the context of the above, most pipes are not bought from mills but from suppliers who have no part in manufacturing the pipe. In such situation, not much is in the control of the purchaser except to keep looking for whatever is available in market and to try to best match their need with whatever is available. That is not best practice though it is the most frequently followed course of action.

8.7 PIPES FROM OTHER SPECIFICATIONS

Pipe from other specifications, especially ASTM or ASME, are commonly used for piping systems particularly in small diameter format and some as larger diameter above ground piping in plants and facilities. These are often ASTM A-106, ASTM A-53 or ASTM A-333 pipes and equipment made from ASTM A 516 various grades, ASTM A 537 Cl I, II, and III plates, etc. When a switch is made from API 5L pipes to ASTM pipes, engineering must evaluate the substituted material based on the SMYS, heat treatment conditions, required impact testing—and both temperature and energy abortion CVN values. For low-temperature service as in arctic regions, the importance of the Charpy impact test temperature and values cannot be overstated with regard to the selection of materials; it must be kept in mind that proven low-temperature grades of material are specified. It may be noted that some ASTM pipes and materials are not mandated to be impact tested, and when required, the impact energy CVN absorbed values for several ASTM/ASME materials are significantly lower than the minimum average 20 ft lbs specified for API 5L pipes.

CHAPTER 9

Fittings and Forgings

Fittings are an essential part of any engineering construction project; pipelines and piping are no exception. The term may cover elbows, tees, reducers (both eccentric and concentric), segmentable bends, induction bends, field bends, flanges, weldolets, anchor forgings and forged fittings. These materials cover a large list of shapes and sizes and they also come in different material strength levels to match pipe material and service conditions. Various grades within the ASTM A-234 specification are specified for a number of wrought carbon and alloy steel fittings for piping. The general properties of various grades are given in Table 9.1 (for reference only. It is strongly recommended that engineers and professionals procure the latest version of the referenced codes and specifications for more current and accurate information on materials).

A number of ASME and Manufacturers Standard Society (MSS) specifications that are commonly used in pipeline construction are discussed, with general information that is useful in making correct decisions in the selection of material for specific project needs. MSS SP 75 Specification for High-Test, Wrought, Butt-Welding Fittings and MSS SP 44 to "steel pipeline flanges"; these are design and material specifications dealing with chemical and mechanical properties of material but also reference several other ASTM specification such as ASTM A-105, A-106, A-53, A-234, A-420, A-694. Some of the specified grades are suitable for low temperature service, for example in arctic regions. Both MSS SP 75 and MSS SP 44 include material with higher yield strength as shown in Table 9.2.

It may be noted that ASME A 234 described earlier is essentially a material specification, and specific fitting dimension and tolerances are governed by one of the ASME B16 specifications discussed later.

Flanges and fittings up to 24 in. in diameter are commonly covered under ASME B16.5 or 16.9, when the diameter exceeds 24 in. it may be necessary to refer to MSS SP 75 for fittings and MSS SP 44 for flanges; MSS specifications also cover alloy steel high-yield strength material. Some of these materials are not ASME material and care must be taken

Table 9.1 Chemical Compositions of Various Wrought Steel Fittings

The Carbon Steel Grades

Grade	C	Mn	P	S	Si	Cr	Mo	Cu	Tensile × 1000 psi	Yield × 1000 psi
WPB	0.30 max.	0.29–1.06	0.050 max.	0.058 max.	0.10 min.				60–85	35
WPC	0.35 max.	0.29–1.06	0.050 max.	0.058 max.	0.10 min.				70–95	40

The Alloy Steel Grades

Grade	C	Mn	P	S	Si	Cr	Mo	Ni	Cu	Tensile × 1000 psi	Yield × 1000 psi
WP1	0.28	0.30–0.90	0.045	0.045	0.10–0.50		0.44–0.65			55–80	30
WP12 Cl 1 and Cl 2	0.050–0.20	0.30–0.80	0.045	0.045	0.60	0.80–1.25	0.44–065			70–95	40
WP11 Cl 1	0.05–0.15	0.30–0.60	0.030	0.030	0.50–1.00	1.00–1.50	0.44–0.65			60–85	30
WP11 Cl 2	0.05–0.20	0.30–0.80	0.04	0.04	0.50–1.00	1.00–1.50	0.44–0.65			70–95	40
WP11 Cl 3	0.05–0.20	0.30–0.80	0.04	0.04	0.50–1.00	1.00–1.50	0.44–0.65			75–100	45
WP22 Cl 1	0.05–0.15	0.30–0.60	0.04	0.04	0.50	1.90–2.60	0.87–1.13			60–85	30
WP22 Cl 3	0.05–0.15	0.30–0.60	0.04	0.04	0.50	1.90–2.60	0.87–1.13			75–100	45
WP 5	0.15	0.30–0.60	0.40	0.30	0.50	4.0–6.0	0.44–0.65			60–85	30
WP 9	0.15	0.30–0.60	0.03	0.03	0.25–1.00	8.0–10.00	0.90–1.10			60–85	30
WPR	0.20	0.40–1.06	0.045	0.050				1.60–2.24	0.75–1.25	63–88	46
WP 91	0.08–0.12	0.30–0.60	0.20	0.10	0.20–0.50	8.0–9.5	0.85–1.05	0.40	V, Co, N, Al	85–110	60

Single numbers are maximum values unless stated otherwise.

Table 9.2 Properties of Forging and Fitting Material Grades

Class Symbol— MSS SP 75	Minimum Yield Strength (psi)	Minimum Tensile Strength (psi) for All Thickness	Minimum Elongation in 2 in. %	Maximum Carbon Equivalent (CEq)$_{IIW}$
WPHY-42	42,000	60,000	25	0.45
WPHY-46	46,000	63,000	25	0.45
WPHY-52	52,000	66,000	25	0.45
WPHY-56	56,000	71,000	20	0.45
WPHY-60	60,000	75,000	20	0.45
WPHY-65	65,000	77,000	20	0.45
WPHY-70	70,000	82,000	18	0.45
MSS SP 44				
Grade MSS SP 44	Minimum Yield Strength (psi)	Minimum Tensile Strength (psi) for All Thickness	Minimum Elongation In 2 inch %	Maximum Carbon Equivalent (CEq)$_{IIW}$
F36	36,000	60,000	20	0.48
F42	42,000	60,000	20	0.48
F46	46,000	60,000	20	0.48
F48	48,000	62,000	20	0.48
F50	50,000	64,000	20	0.48
F52	52,000	66,000	20	0.48
F56	56,000	68,000	20	0.48
F60	60,000	75,000	20	0.48
F65	65,000	77,000	18	0.48
F70	70,000	80,000	18	0.48

to ensure that such material meets the necessary calculations. The material specified also differs in some grades of MSS SP products, and also address the low temperature and impact energy absorption requirements. The fittings applicable are often quenched and temper heat treated.

The term forgings for this discussion include flanges, forged fittings, anchor forgings, and weldolets. These are heavy materials with varied microstructure, which behave very differently in their cooling and heating cycles, and this gives rise to complexities when they are welded with other wrought material. Specifically in welding, the bulk of material works as a heat sink affecting the directional heat flow during welding. Due to the thickness of the material, the heat flow is often three-directional, thus raising the relative thickness values. This gives rise to the situation where simple and uniform cooling cannot be predicted, and the carbon equivalent (CEq) value of forgings is often higher than the CEq values specified for pipes that are welded. This demands that when welding pipe to a forging, a suitable welding procedure is developed to control cooling rates to avoid formation of harmful martensite and subsequent cracking. The higher alloyed forging specifications like ASTM A-694 and MSS SP 44 specify the maximum CEq, and these are often higher than for most of the wrought materials they are welded with. Engineers have the responsibility to make an assessment and to determine whether they would specify a lower value as an acceptable maximum CEq to meet the requirements of the work in question; this is especially important in pipeline construction work where welding procedure specification (WPS) are qualified with limited preheating and without postweld heat treatment (PWHT). As a general guidance, a CEq exceeding 0.39 should be avoided if proper preheat, or PWHT, or both cannot be included in the welding procedure. It may be added that CEq is a relative number, not a percentage, though many specifications and books erroneously report CEq as percentage (%).

In the following sections, some related specifications are introduced and discussed briefly; a specification must be reviewed prior to making a decision.

9.1 ASME/ANSI B16.5—PIPE FLANGES AND FLANGED FITTINGS

This Standard for flanges and flanged fittings covers pressure−temperature ratings, materials, dimensions, tolerances, marking, testing, and methods of designating openings for pipe flanges and flanged fittings.

The Standard includes flanges with rating class designations 150, 300, 400, 600, 900, 1500, and 2500 in sizes NPS 1/2 through NPS 24. The requirements are given in both metric and US units. The Standard is limited to flanges and flanged fittings made from cast or forged materials, and blind flanges and certain reducing flanges made from cast, forged, or plate materials. Also included in this Standard are requirements and recommendations regarding flange bolting, flange gaskets, and flange joints.

Flanges have temperature and pressure class ratings assigned to them. In Table 9.3 values of the maximum non-shock pressure (psig)

Temperature (°F)	Pressure Class (lb)						
	150	300	400	600	900	1500	2500
	Hydrostatic Test Pressure (psig)						
	450	1125	1500	2225	3350	5575	9275
−20 to 100	285	740	990	1480	2220	3705	6170
200	260	675	900	1350	2025	3375	5625
300	230	655	875	1315	1970	3280	5470
400	200	635	845	1270	1900	3170	5280
500	170	600	800	1200	1795	2995	4990
600	140	550	730	1095	1640	2735	4560
650	125	535	715	1075	1610	2685	4475
700	110	535	710	1065	1600	2665	4440
750	95	505	670	1010	1510	2520	4200
800	80	410	550	825	1235	2060	3430
850	65	270	355	535	805	1340	2230
900	50	170	230	345	515	860	1430
950	35	105	140	205	310	515	860
1000	20	50	70	105	155	260	430

Table 9.3 Maximum Allowable Non-Shock Pressure (psig)

for pressure class ratings of 150–2500 are given (for reference only). The dimensions for various classes ranging from Class 150 to Class 2500 ASME B16.5 are given and readers are advised to reference these specifications for up to date correct information.

9.2 ASME/ANSI B16.9—FACTORY-MADE WROUGHT STEEL BUTT WELDING FITTINGS

This Standard covers overall dimensions, tolerances, ratings, testing, and markings for wrought factory-made butt welding fittings in sizes NPS 1/2 through 48 (DN 15 through 1200). Various shapes and sizes of fittings are manufactured to meet all possible engineering requirements and they are available in different class ratings or wall thickness classifications. The dimensions of various types of fittings listed below are given in ASME B16.9 and readers are encouraged to reference the tables therein for data on various critical dimensions for each of the following:

- Long radius elbows
- Long radius reducing elbows

- Long radius returns (U-bends)
- Short radius elbows
- Long radius returns
- 3D-90° and 45° elbows
- Tees and crosses
- Lap joint stub ends
- Caps
- Eccentric and concentric reducers.

The tolerances applying to the dimensions of the fittings listed above are detailed in a table in ASME B16.9.

9.3 ASME/ANSI B16.11—FORGED STEEL FITTINGS, SOCKET WELDING AND THREADED CONNECTIONS

This Standard covers ratings, dimensions, tolerances, marking, and material requirements for forged fittings, both socket welding and threaded connections.

9.4 ASME/ANSI B16.14—FERROUS PIPE PLUGS, BUSHINGS AND LOCKNUTS WITH PIPE THREADS

This Standard for ferrous pipe plugs, bushings, and locknuts with pipe threads addresses:

a. Pressure—temperature ratings
b. Size
c. Marking
d. Materials
e. Dimensions and tolerances
f. Threading
g. Pattern taper.

9.5 ASME/ANSI B16.20—METALLIC GASKETS FOR PIPE FLANGES RING-JOINT, SPIRAL-WOUND, AND JACKETED

This Standard covers materials, dimensions, tolerances, and markings for metal ring-joint gaskets, spiral-wound metal gaskets, and metal-jacketed gaskets and filler material. These gaskets are dimensionally suitable for use with flanges described in the reference flange standards

ASME/ANSI B16.5, ASME B16.47, and API-6A. This Standard covers spiral-wound metal gaskets and metal-jacketed gaskets for use with raised face and flat face flanges.

9.6 ASME/ANSI B16.21—NONMETALLIC FLAT GASKETS FOR PIPE FLANGES

This Standard for nonmetallic flat gaskets for bolted flanged joints in piping includes the following:

a. Types and sizes
b. Materials
c. Dimensions and allowable tolerances.

9.7 ASME/ANSI B16.25—BUTT WELDING ENDS

This Standard covers the preparation of butt welding ends of piping components to be joined into a piping system by welding. It includes requirements for welding bevels, for external and internal shaping of heavy-wall components, and for preparation of internal ends (including dimensions and tolerances). Coverage includes preparation for joints with the following:

a. No backing rings
b. Split or noncontinuous backing rings
c. Solid or continuous backing rings
d. Consumable insert-rings
e. Gas tungsten are welding (GTAW) of the root pass. Details of preparation for any backing ring must be specified in ordering the component.

9.8 ASME/ANSI B16.28—WROUGHT STEEL BUTT WELDING SHORT RADIUS ELBOWS AND RETURNS

This Standard covers ratings, overall dimensions, testing, tolerances, and markings for wrought carbon and alloy steel butt welding short radius elbows and returns. The term wrought denotes fittings made of pipe, tubing, plate, or forgings.

9.9 ASME/ANSI B16.36—ORIFICE FLANGES

This Standard covers flanges (similar to those covered in ASME B16.5) that have orifice pressure differential connections. Coverage is limited to the following:

a. Welding neck flanges Classes 300, 400, 600, 900, 1500, and 2500
b. Slip-on and threaded Class 300
c. Orifice, nozzle, and Venturi flow rate meters.

9.10 ASME/ANSI B16.39—MALLEABLE IRON THREADED PIPE UNIONS

This Standard for threaded malleable iron unions, classes 150, 250, and 300, provides requirements for the following:

a. Design
b. Pressure—temperature ratings
c. Size
d. Marking
e. Materials
f. Joints and seats
g. Threads
h. Hydrostatic strength
i. Tensile strength
j. Air pressure test
k. Sampling
l. Coatings
m. Dimensions.

9.11 ASME/ANSI B16.42—DUCTILE IRON PIPE FLANGES AND FLANGED FITTINGS, CLASSES 150 AND 300

This Standard covers minimum requirements for classes 150 and 300 cast ductile iron pipe flanges and flanged fittings. The requirements covered are as follows:

a. Pressure—temperature ratings
b. Sizes and method of designating openings
c. Marking

d. Materials
e. Dimensions and tolerances
f. Blots, nuts, and gaskets
g. Tests.

9.12 ASME/ANSI B16.47—LARGE DIAMETER STEEL FLANGES: NPS 26 THROUGH NPS 60

This Standard covers pressure–temperature ratings, materials, dimensions, tolerances, marking, and testing for pipe flanges in sizes NPS 26 through NPS 60 and in ratings classes 75, 150, 300, 400, 600, and 900. Flanges may be cast, forged, or plate (for blind flanges only) materials. Requirements and recommendations regarding bolting and gaskets are also included.

9.13 ASME/ANSI B16.48—STEEL LINE BLANKS

This Standard covers pressure–temperature ratings, materials, dimensions, tolerances, marking, and testing for operating line blanks in sizes NPS 1/2 through NPS 24 for installation between ASME B16.5 flanges in the 150, 300, 600, 900, 1500, and 2500 pressure classes.

9.14 ASME/ANSI B16.49—FACTORY-MADE WROUGHT STEEL BUTT WELDING INDUCTION BENDS FOR TRANSPORTATION AND DISTRIBUTION SYSTEMS

This Standard covers design, material, manufacturing, testing, marking, and inspection requirements for factory-made pipeline bends of carbon steel materials having controlled chemistry and mechanical properties, produced by the induction bending process, with or without tangents. This Standard covers induction bends for transportation and distribution piping applications (e.g., ASME B31.4, B31.8, and B31.11). Process and power piping have differing requirements and materials that may not be appropriate for the restrictions and examinations described herein, are therefore not included in this Standard.

CHAPTER 10

Valves

Valves in oil and gas installations are governed by several standards and specifications that are issued by many organizations. They are dynamic documents that reflect sound engineering and changes in the needs of the industry based on engineering practice, changes in market demands, and changes in technology and manufacturing procedures.

Valve standards focus on important aspects that include types of valve, for example gate, globe, or check valves, and their constituent materials.

All aspects of valve design, functionality, inspection, and testing are covered in dozens of ASME, API, and MSS documents and various other international standards. The number of codes, standards, and specifications can render the procurement of valves and associated products a job for experts only.

A good understanding of the primary standards relating to these products is essential. It is recommended that all current valve standards that apply to the project are referenced.

In the following sections, a brief introduction to some of the specifications commonly in use is given.

10.1 API 598 VALVE INSPECTION AND TESTING

This specification covers testing and inspection requirements of gate, globe, ball, check, plug, and butterfly valves. The test pressure for each valve must be determined on the basis of the tables given in ASME/ANSI B16.34.

10.2 API 600 STEEL VALVES—FLANGED AND BUTT WELDED ENDS

API 600 and ISO 10434 are the primary steel gate valve specifications. They address design construction and testing criteria. An appendix in API 600 addresses pressure seal valves. API 600 also lists important dimensions such as stem diameter minimums, wall thickness, and stuffing box size.

Another important gate valve specification is ASME/ANSI B16.34. This document gives extensive details on valves built to meet ASME boiler code pressure–temperature ratings. One important point of difference from API 600 in relation to ASME/ANSI B16.34 is the wall thickness requirement. API 600 requires a heavier wall for a given pressure rating than does ASME/ANSI B16.34. This difference must be kept in mind when working with two different codes.

10.3 API 602: COMPACT STEEL GATE VALVES—FLANGED, THREADED, WELDING, AND EXTENDED BODY ENDS

This specification addresses gate valve sizes of 4 in. and below. Valves from Class 150 through Class 1500 are covered by API 602. This specification covers the same details for small forged gate valves that API 600 does for larger valves. API 602 further gives dimensions for extended body valves which are used extensively in industrial facilities.

Similarly to API 600, API 602 requires a heavier wall for classes 150, 300, and 600 as compared with ASME/ANSI B16.34 requirements.

10.4 API 603 CLASS 150 CAST, CORROSION RESISTANT FLANGED END GATE VALVES

This standard covers corrosion-resistant bolted bonnet gate valves with flanged or butt welded ends in sizes NPS ½–24 in., corresponding to nominal pipe sizes ASME B31.10M and classes 150, 300, and 600 as specified in ASME B16.34.

10.5 API 608 METAL BALL VALVES—FLANGED AND BUTT WELDED ENDS

This specification covers classes 150 and 300 metal ball valves that have either butt weld or flanged ends and are for use in on-off service. It addresses the design and construction criteria. The important feature to note is that the working pressure for ball valves must be based on the seat material and not on the class of the valve.

These valves are commonly available in cast steel meeting ASTM A351 grade CF8M and CF8; however, other corrosion-resistant alloys meeting other specifications and grades are also used to manufacture these valves.

10.6 API 609: BUTTERFLY VALVES—DOUBLE FLANGED, LUG AND WAFER TYPE

These valves are designed to meet up to class 600 rating and are intended to be fitted between flanges. The standard covers design, materials, face-to-face dimensions, pressure–temperature ratings, and examination, inspection, and test requirements for gray iron, ductile iron, bronze, steel, nickel-base alloy, or special alloy butterfly valves that provide tight shutoff in the closed position and are suitable for flow regulation.

10.7 TESTING OF VALVES

As described earlier, the test specification for most valves is API 598 "Valve Inspection and Test." Most metallic seated valves larger than ANSI size 2 in. have an allowable leakage rate that is listed in API 598. Some valve types such as bronze gate, globe, and check valves are usually not tested using API 598. These are normally tested according to MSS SP 61 "Pressure Testing of Steel Valves."

Pipeline valves are often specified to meet API 6D/ISO14313 requirements. This document covers the design, materials, and dimensions of valves for pipeline service. The testing requirements of API 6D differ from those of API 598. The primary difference is API 6D's zero allowable leakage on closure tests. Since most of the valves built to API 6D are resilient seated, this is easily achieved. However, when the test standard is applied to metallic seated wedge gate, globe, or check valves, compliance can be difficult.

10.8 NACE TRIM AND NACE MATERIAL

10.8.1 NACE MR0175/ISO 15156 Standard Material Requirements for Sulfide Stress Cracking Resistant Metallic Materials for Oilfield Equipment

This Standard is now an ANSI specification; it is no longer merely a recommended practice as was the case in the recent past. MR0175 is divided into three parts and helps to determine the severity level of difficult environments in relation to various materials. The specification does not list materials but gives guidance on how to make assessments and, if required, to qualify materials for specific service conditions. Most of the materials are identified by their UNS numbers and fabrication techniques.

In valve manufacturing, often the term "NACE" trim is used; this simply means that the trim material shall be compliant with NACE/ANSI MR0175/ISO 15156 and meet the given service conditions as specified. Compliance with NACE lessens the likelihood of H_2S-induced cracking.

10.9 ASME CODES AND OTHER SPECIFICATIONS

Although ASME Section IX Welding and Brazing, ASME Section VII Div 1 Pressure Vessels Design, ASME B31.8 Gas Transmission and Distribution Piping Systems, and ASME B31.4 Pipeline Transportation System for Liquid Hydrocarbon and Other Liquids are not valve standards, they have a strong influence on valve design, manufacturing, testing, and application. In relation to pipeline, these specifications are regularly consulted for welding, design, manufacture, and application.

Several international organizations have also issued valve standards, including British Standards (BS), International Standards Organization (ISO), and the Canadian Standards Association (CSA). Some of the relevant valve standards are BS 1873 and BS 5352 for globe valves, BS 6364 for cryogenic service valves, and BS 1868 and BS 5352 for steel check valves. These documents are excellent starting points for persons needing guidance in these particular areas (Table 10.1).

10.9.1 Valve Materials
10.9.1.1 Trim Material
The internal metal parts of a valve, such as the ball, stem, and metal seats or seat retainers, should be of the same nominal chemical composition as the shell and have mechanical and corrosion-resistance properties similar to those of the shell. The purchaser may specify a higher quality trim material.

Common pipeline material for specific parts is given in Tables 10.2–10.5.

Table 10.1 Common Valve Testing Standards

Specification	Valve Types
API 598	All
API 6D	Pipeline
ANSI B16.34	All
MSS SP 61	All
MSS SP 67	Butterfly
MSS SP 68	Butterfly
MSS SP 70	Iron
MSS SP 80	Bronze
MSS SP 81	Stainless steel knife gate valves

Table 10.2 Ball Valve Material—Body and Tailpiece Material

Material Specification	Material	Service Condition
ASTM A105	Carbon steel	Good for normal pipeline working conditions. However, it should be noted that upon prolonged exposure to temperatures above 427°C, the carbide phase of carbon steel may be converted to graphite.
ASTM A352 LF2	Low temperature	For service from −46°C to 340°C. This material must be quenched and tempered to obtain tensile and impact properties needed at subzero temperatures.
ASTM A182 F304	18Cr−8Ni	Good creep strength, and corrosion and oxidation resistance when exposed to temperatures above 427°C.
ASTM A182 F316	18Cr−10Ni−2Mo	Good creep strength, and corrosion and oxidation resistance when exposed to temperatures above 427°C, and it is resistant to formulation of Σ phase.

Table 10.3 Fasteners Material

	Normal Service	NACE Compliant	Low Temperature	Corrosion Resistant	Others
Stud	Al93 Gr. B7	Al93 Gr. B7M	A320 Gr.L7	A193 Gr.B8(M)	Monel
Nut	Al94 Gr.2H	Al94 Gr. 2HM	Al 94 Gr.L4	Al94 Gr. 8(M)	Monel

Component materials for NACE standards are given in Tables 10.6−10.8.

There are several other valve specifications that address specific needs.

Table 10.4 Stem Materials

ASTM Specification	Service Condition
A276 420/410	Good for service up to 425°C where corrosion and oxidation are not factors.
A105 ENP	Good for service up to 425°C where corrosion and oxidation are not factors.
A747 17-4 PH	Very high tensile material. Often used when differential hardness is required due to its resistance to galling. Material has higher corrosion resistance compared to straight chromium alloy steels.
A182 F316	Good creep resistance, corrosion and oxidation resistance when exposed to temperatures above 427°C, and it is resistant to formulation of Σ phase.

Table 10.5 Ball Materials

ASTM Specification	Material	Service Condition
A105(N) ENP	Carbon steel	For service up to 538°C where corrosion and oxidation are not a factor.
A182 F304(L), A351 CF8(3)	18Cr−8Ni	Good creep resistance properties, and corrosion and oxidation resistance when exposed to temperatures above 427°C.
A182 F316 A351 CF8M	18Cr−10Ni−2Mo	Good creep resistance properties, and corrosion and oxidation resistance when exposed to temperatures above 427°C, and it is resistant to formation of Σ phase.
A182 F316 A351 CF3M	18Cr−10Ni−2Mo	Good creep resistance properties, corrosion and oxidation resistance when exposed to temperatures above 427°C, and it is resistant to formation of Σ phase.

Table 10.6 NACE Compliance Body and Tailpiece Materials

Material Specification	Type of Material	Application
ASTM A105	Forged carbon steel with ENP coating	For valve sizes 2 in. and above for all NACE compliant valves. The ENP (Electroless Nickel Plated) thickness is often limited to 0.003 in.
A182 F304/ F316	Stainless steel forging	For valve sizes 2 in. and above for all NACE compliant valves.

10.10 API 6D SPECIFICATION FOR PIPELINE VALVES

This API specification is possibly the most referenced valve specification in the pipeline industry. API Specification 6D is an adoption from ISO14313:1999, Petroleum and Natural Gas Industries—Pipeline Transportation Systems—Pipeline Valves. This International Standard

Table 10.7 NACE Compliance Ball Materials

Material Specification	Type of Material	Application
ASTM A105	Forged carbon steel with ENP coating	For valve sizes 2 in. and above for all NACE compliant valves. ENP (Electroless Nickel Plated) thickness is often limited to 0.003-inch
A182 F304/ F316	Stainless steel forging	For valve sizes 2 in. and above for all NACE compliant valves.

Table 10.8 NACE Compliance Stem Materials

Material Type	Properties	Application
Carbon steel bar	Maximum hardness not to exceed 22 HRC	For all classes of valves of size 2–4 in. Some larger sizes apply above 6 in.
Stainless steel bar	17-4 PH	For all corrosion service or low temperature service requirements.

specifies requirements and gives recommendations for the design, manufacturing, testing, and documentation of ball, check, gate, and plug valves for application in pipeline systems.

Types of valves covered under this specification are gate valves, lubricated and nonlubricated plug valves, ball valves, and check valves with full or reduced opening configurations. The valves covered are of class ranges 150 (PN20) to 2500 (PN 420).

10.11 API 6FA

This specification establishes the requirements for testing and evaluating the pressure containing performance of API 6D valves and also wellhead Christmas-tree equipment according to API 6A. The performance requirements are established regardless of the pressure rating or size.

The test covers the requirements of leakage through the valve and also the external leakage after exposure to fire reaching temperatures between 1400°F and 1800°F (761°C and 980°C) for 30 min.

10.12 API 526 FLANGED STEEL PRESSURE RELIEF VALVES

This Standard addresses the basic requirements for direct spring-loaded pressure relief valves and pilot-operated pressure relief valves. This includes the orifice designation and area; valve size and pressure rating, inlet and outlet; materials; pressure—temperature limits; and center-to-face dimensions, inlet and outlet.

10.13 API 527 SEAT TIGHTNESS OF PRESSURE RELIEF VALVES (2002)

This API Specification describes methods of determining the seat tightness of metal and soft-seated pressure relief valves, including those of conventional, bellows, and pilot-operated designs.

10.14 ANSI/API STD 594 CHECK VALVES: FLANGED, LUG, WAFER, AND BUTT WELDING

API Standard 594 covers design, material, face-to-face dimensions, pressure—temperature ratings, and examination, inspection, and test requirements for two types of check valves.

10.15 ANSI/API 599: METAL PLUG VALVES—FLANGED, THREADED, AND WELDING ENDS

A purchasing specification that covers requirements for metal plug valves with flanged or butt welding ends, and ductile iron plug valves with flanged ends, in sizes NPS 1 through NPS 24, which correspond to nominal pipe sizes in ASME B36.10 M. Valve bodies conforming to ASME B16.34 one side may have flanged end and the other end is the welding end. It also covers both lubricated and nonlubricated valves that have two-way coaxial ports and includes requirements for valves fitted with internal body, plug, or port linings or applied hard facings on the body, body ports, plug, or plug port.

10.16 ASME/ANSI B16.38—LARGE METALLIC VALVES FOR GAS DISTRIBUTION

This Standard covers only manually operated metallic valves in nominal pipe sizes 2.5—12 in. that have an inlet and outlet on a common centerline. These valves are suitable for controlling the flow of gas from open to fully closed. These valves are often used in distribution service lines, where the maximum gauge pressure at which such distribution piping systems may be operated is directed by regulatory bodies such as the US Code of Federal Regulations, Title 49, Part 192, Transportation of Natural and Other Gas by Pipeline. The regulated maximum pressure in the United States is 125 psi (8.6 bar). Valve seats, seals, and stem packing may be nonmetallic.

10.17 ASME/ANSI B16.33—MANUALLY OPERATED METALLIC GAS VALVES FOR USE IN GAS PIPING SYSTEMS UP TO 125 psig

This Standard covers requirements for manually operated metallic valves sizes NPS 1.2 through NPS 2, for outdoor installation as gas shut-off valves at the end of a gas service line and in advance of the gas regulator and meter where the designated gauge pressure of the gas piping system does not exceed 125 psi (8.6 bar). The Standard applies to valves operating in a temperature environment between $-20°F$ and $150°F$ ($-29°C$ and $66°C$).

This Standard sets forth the minimum capabilities, characteristics, and properties which a valve at the time of manufacture must possess in order to be considered suitable for use in gas piping systems.

10.18 ASME/ANSI B16.40—MANUALLY OPERATED THERMOPLASTIC GAS VALVES

This Standard covers manually operated thermoplastic valves in nominal sizes 1.2—6 in. These valves are suitable for use below ground in thermoplastic distribution mains and service lines. In the United States, the maximum pressure at which such distribution piping systems may be operated must be in accordance with the Code of Federal Regulation, Title 49, Part 192, Transportation of Natural and Other Gas by Pipeline; Minimum Safety Standards, for temperature ranges of $-20°F$ to $100°F$ ($-29°C$ to $38°C$). This Standard sets qualification requirements for each nominal valve size for each valve design. It sets requirements for newly manufactured valves for use in below-ground piping systems for natural gas, including synthetic natural gas (SNG), and liquefied petroleum (LP) gas distributed as vapor, with or without the admixture of air, or as a mixture of gases.

10.19 ASME/ANSI B16.10—FACE-TO-FACE AND END-TO-END DIMENSIONS OF VALVES

This Standard covers face-to-face and end-to-end dimensions of straightway valves, and center-to-face and center-to-end dimensions of angle valves. Its purpose is to assure installation interchangeability for valves of a given material, type size, rating class, and end connection.

10.20 ASME/ANSI B16.34—VALVES—FLANGED, THREADED, AND WELDING END

This Standard applies to new valve construction and covers pressure—temperature ratings, dimensions, tolerances, materials, nondestructive examination requirements, testing, and marking. The specification covers cast, forged, and fabricated material with flanged, threaded, and weld-ends, and wafer or flangeless valves of steel, nickel-base and other alloys. Wafer or flangeless valves, bolted or through-bolt types, that are installed between flanges or against a flange are treated as flanged end valves.

INDEX

A

AISI, 86
Alaska pipeline, 4–5, 25
Anchor Block, 19
Anisotropy, 91–92
API 5L, 83–86, 92–98
Arctic pipeline, 3–4, 37
Arctic soil, 22
ASME B 31.8, 85–86, 112

B

Beaufort Sea, 4, 7
Berger Commission, 5–6
Bridges, 33
Brittle fracture, 78
Buoyancy, 17–19, 33

C

Canadian Standard Association Specification, 36–37
Carbon, 24
Carbon equivalent (CEq), 101–102
Caspian Sea, 6–7, 31
Cathodic protection, 28
Charpy, 82–83, 92, 97–98
Chuchki Sea, 7
Cleavage fracture, 78
Cold spring, 23
Communication, 49, 53, 55, 60–61
Compressive strain capacity, 23
Compressors, 38
Controlled rolling, 91
Crack arrest theory, 79
Crack initiation, 78
Crack propagation, 78
Crack-arrest temperature, 81
Cryogenic service, 79–80
CSA Z662, 36–37
CTOD, 74, 79

D

Deepwater Horizon, 7
Deflections, 19
Deformation, 91

Design

Design, 3–4, 10, 22–26, 31, 35–38, 47, 51–56, 73–75, 83, 85, 87, 92, 94, 99, 107, 109–112, 114–118
Drake Gas Field, 32
DSAW, 93–94
Ductile fracture, 78
Ductility, 74–77
Duty cycle, 65–66
DWTT, 85, 96

E

Electric motors, 64–66
Electrical equipment, 63–68
Energy absorption, 83–84, 86–87, 99–101
EPRG, 86
Equilibrium, 21
ERW, 92–93
European Pipeline Research Group, 86
European Union Directives, 47
Excavation, 37, 46–47
Expansion and flexibility, 15–16

F

Failure mechanism, 78
Fall protection, 46
Fiber textures, 91
Flow assurance, 25
Formability, 90
Fracture Mechanics, 77–79
Free span support, 23

G

GPS locations, 24–25
Grain refinement, 91
Gulf of México, 7

H

Hardness, 74
HAZ, 93
HFI, 93
HFW, 92–93
Hoop stress, 23–24
HSLA steels, 24

I

Ice gouging, 31
Intergranular fracture, 78
ISO 3183, 85, 92, 96–97
Izod, 82–83

K

KIC, 74

L

Limit states design, 22
Linear coefficient of expansion, 24
Load inertia, 64–65
Load speed, 64–65
Longitudinal expansion, 34
Longitudinal strain, 22

M

Mackenzie Delta, 5
Mackenzie Valley, 5–6
Manufacturers Standard Society (MSS), 99
Mesnager, 82–83
Microstructure, 74, 90–91, 101–102
Microwave, 49
Modulus of elasticity, 24
Moment of inertia, 16

N

National Electrical Manufacturers Association
  (NEMA), 61, 63, 65–68
Nil Ductility Temperature, 79
North Slope, 5
NorthStar, 32

O

OSHA, 45–47

P

Permafrost, 25–28
Pig launching, 33
Pipeline and facilities, 10
Pipeline crossings, 32–33
Plastic deformation, 22, 77, 84
Poisson's ratio, 19, 24
PSL2, 87, 92, 97
Pump stations, 4–5, 38
Pump/compressor stations, 33
Recrystallization, 91
River and waterways crossings, 11

S

Safety, 38, 43, 47, 117–118
Satellite, 49, 55
SAW, 93–94
SAWH, 93–94
SAWL, 93–94
SCADA, 49–57
Service factors, 66
Shear fracture, 78
Siberia, 3, 30
SMYS, 24, 92, 94
Spiral welded pipe, 91–92
Strain Based, 22–25
Stresses in the pipe, 19
Subsea Arctic pipelines, 31–32

T

Tensile strain capacity, 23
Tensile strength, 74, 82, 106
Thaw settlement, 28–29
Thawing, 4, 26–28
Thermal coefficient of expansion, 19
Thermal expansion, 34–35, 82
Thermal stresses, 35
Toughness, 74, 77–78
Transformation, 91
Transition temperature, 84
Trim Material, 112
Tundra, 7, 26

U

Underwriters' Laboratories, 63
Unsupported span of pipe, 16–17

V

Valve Materials, 112–113
Valves, 33, 109–112, 114t, 115t, 116–118

W

Wildlife crossing, 32–33

Y

Yamal peninsula, 30
Yield strength, 74
Yoloten-Osman, 6
Young's Module, 19

Z

Zigzag configuration, 34

www.ingramcontent.com/pod-product-compliance
Lightning Source LLC
Chambersburg PA
CBHW070735220326
41598CB00024BA/3436